NASA/SP-2010-577

Evaluation of the NASA Arc Jet Capabilities to Support Mission Requirements

Office of the Chief Engineer

May 2010

Preface

NASA accomplishes its strategic goals through human and robotic exploration missions. Many of these missions require launching and landing or returning spacecraft with human or return samples through Earth's and other planetary atmospheres. Spacecraft entering an atmosphere are subjected to extreme aerothermal loads. Protecting against these extreme loads is a critical element of spacecraft design. The safety and success of the planned mission is a prime concern for the Agency, and risk mitigation requires the knowledgeable use of thermal protection systems to successfully withstand the high-energy states imposed on the vehicle. Arc jets provide ground-based testing for development and flight validation of re-entry vehicle thermal protection materials and are a critical capability and core competency of NASA.

The Agency's primary hypersonic thermal testing capability resides at the Ames Research Center and the Johnson Space Center and was developed and built in the 1960s and 1970s. This capability was critical to the success of Apollo, Shuttle, Pioneer, Galileo, Mars Pathfinder, and Orion. But the capability and the infrastructure are beyond their design lives. The complexes urgently need strategic attention and investment to meet the future needs of the Agency.

The Office of Chief Engineer (OCE) chartered the Arc Jet Evaluation Working Group (AJEWG), a team of experienced individuals from across the Nation, to capture perspectives and requirements from the arc jet user community and from the community that operates and maintains this capability and capacity. This report offers the AJEWG's findings and conclusions that are intended to inform the discussion surrounding potential strategic technical and investment strategies. The AJEWG was directed to employ a 30-year Agency-level view so that near-term issues did not cloud the findings and conclusions and did not dominate or limit any of the strategic options.

The OCE would like to thank the members of the user community who gave presentations, interviews, responded to questions, and offered their vision of the future. The OCE thanks the facility managers and personnel who hosted the AJEWG at their Center, conducted tours of their arc jet complexes, and gave presentations, answered questions, and offered their vision of the future. The OCE also extends its appreciation to the AJEWG for their time and effort in developing this report: Anthony Calomino, Lead, GRC; Walt Bruce, LaRC; Peter Gage, Neerim Corp.; Dennis Horn, AEDC (Retired); Mike Mastaler, HQ; Don Rigali, DOE (Retired); Judee Robey, HQ; Linda Voss, Dell-Perot; Jerry Wahlberg, LaRC and NC State (Retired); and Calvin Williams, HQ.

Mike Ryschkewitsch
May 4, 2010

Executive Summary

The Arc Jet Evaluation Working Group (AJEWG) was chartered by the NASA Office of the Chief Engineer to provide an independent technical evaluation of NASA arc jet testing capability considering existing test complexes and the need to support planned and future NASA mission requirements. Future mission classes considered include human return from low Earth orbit, human return from beyond low Earth orbit, both human and robotic exploration of Mars, and deep space robotic exploration and Earth return. Arc jet testing support for all mission phases, from precursor technology maturation through sustaining engineering of operational systems, was to be included in the evaluation. The AJEWG considered the availability and use of arc jet test facilities, focusing on the Johnson Space Center, the Ames Research Center, the Arnold Engineering Development Center, and the Boeing Large Core Arc Tunnel, and generated findings and conclusions to inform long-term investment strategies for NASA. The AJEWG considered use of foreign facilities unacceptable due to restrictions imposed by International Traffic in Arms Regulations, security, and lack of control of facility access. The Nation could not preserve its leadership in the development and modeling of thermal protection materials if it depends on the arc jet testing facilities of a foreign country. The AJEWG findings and conclusions are derived from a variety of resources, including a review of maintenance plans, site visits, reference documents, and invited presentations from the user community.

The AJEWG determined that proposed NASA missions over the next 30 years will require arc jet capabilities beyond what exists today. Based on evaluation of available information, described in Section 6.0 of the report, the AJEWG concludes that a new build offers functional advantages over upgrading existing aging complexes without appreciable cost penalty when considering a 30-year investment. Furthermore, no current facility, including those at Arnold Engineering Development Center and the Large Core Arc Tunnel, can deliver the heating rates, pressures, and shear levels at the scale and duration needed for cost-effective, weight-efficient, and reliable design of thermal protection systems for safe return from Mars or near-Earth objects.

The AJEWG determined that NASA has a critical and strategic need for arc jet ground test capability to meet its unique mission set, and must make its own investment to support future needs.

The NASA arc jet complexes at Johnson and Ames are aging and major infrastructure investment decisions must be made soon. The condition of the equipment and infrastructure that comprise and support these complexes are typical of research and test systems constructed 30 to 40 years ago. Maintenance and recapitalization have been erratic due to fluctuating project and complex maintenance funding levels, resulting in a mix of old and new equipment. Both complexes have identified a near-term need for boiler upgrades; however, both boilers have been recently inspected and certified safe to operate and will meet emissions requirements for the next 5 years. Based on a review of the existing infrastructure, discussions with complex managers, and the availability metrics, the AJWEG expects that both the Ames and Johnson arc jet complexes are capable of sustaining safe operations for a 5-year period, which provides NASA with a critical opportunity to institute appropriate action to establish an arc jet complex to meet NASA's 30-year needs. Although the older equipment has been adequately maintained for safe operation, the real and increasing risk of a major system failure and extended complex down time highlights the need for immediate action.

Within five years, NASA must be building a single arc jet testing complex with a robust infrastructure and capability to support thermal protection system certification for large-mass Mars entry and safe return to Earth.

NASA must start immediately to gain advocacy and establish funding and design plans for the test complex infrastructure and perform the technology development needed for advanced capability. A design team comprised of NASA, DOD, and industry arc jet experts should be convened and tasked to complete preliminary design review for the test complex within 12 months. The team should identify designs that optimize test efficiency, enhance instrumentation access, and improve diagnostic capability for test models. The design team should consider a number of possible sites and assess current and expected future restrictions on power, water, noise, pollution, and availability of suitable personnel. To optimize operational robustness, the team should also consider the potential for natural disasters that could threaten extended downtime. A detailed cost analysis for the construction and 40-year operation of the complex for all site locations considered should be a primary factor for a final build decision. Once a new capability is established, NASA needs to commit to continuous and consistent institutionalized funding to avoid the adverse effects of erratic program funding for maintenance and upgrades and to maximize value to the Agency.

A focused activity to define complex infrastructure requirements, establish technology enhancements and design plans, and acquire financial support and construction approval should start immediately and be scheduled to support building within the next 5 years.

Operations at existing complexes should be sustained, until a final investment decision has been made and funds obligated, to support current missions and the development of advanced capability for the new build. Mothballing either complex is effectively a decision to close that complex. Sustaining operation at both complexes offers NASA an opportunity to use existing facilities to develop the arc jet technology needed to meet the future capability by exploiting expected short-term reductions in arc jet test demand. During this period, no major infrastructure investment should be made that is not necessary for safe operation. The AJEWG agrees with a cooperative management structure proposed by Johnson and Ames for the two existing complexes, as discussed in Section 5.0. The AJEWG noted that current operations staff is a highly valuable NASA resource. They possess unique knowledge and skills that are integral to successful operation and NASA should maximize the opportunity to retain and transition this staff to the new complex. However, also discussed in Section 5.0, the AJEWG concluded that current staff levels at both NASA complexes could be lower. To reduce operation costs, an independent assessment of minimum staffing levels required for safe and effective operation should be completed at both the Ames and Johnson complexes and appropriate reductions instituted within a year.

NASA should maintain testing capability at both the Johnson and Ames arc jet complexes for the time period required to begin building the new complex. NASA should reduce current staff levels at Johnson and Ames and manage a scheduled phase-out plan that offers an efficient transition to test operation at the new complex.

The AJEWG considers these immediate actions to be positive steps that will transition NASA from having a high-risk reliance on the existing, aged arc jet test infrastructure and capability to a single robust complex with improved and cost-effective capability that meets future critical testing needs for NASA and the Nation for the next 30 years.

Table of Contents

PREFACE 2

EXECUTIVE SUMMARY 3

1.0 INTRODUCTION 9
   1.1 SCOPE 9
   1.2 BACKGROUND 9
   1.3 EVALUATION APPROACH 10
   1.4 CURRENT CHANGING CLIMATE FOR FUNDING AND REQUIREMENTS 11
   1.5 DOCUMENT ORGANIZATION 11

2.0 ENGINEERING USES AND ACCOMPLISHMENTS OF ARC JETS 12
   2.1 MATERIAL RESPONSE MODELING 14
   2.2 COMPUTATIONAL FLUID DYNAMICS MODELING 15
   2.3 MAXIMIZING VALUE OF ARC JET TESTING 17
   2.4 SUMMARY 18

3.0 MISSION/FUTURE REQUIREMENTS/NATIONAL NEEDS 19
   3.1 RETURN FROM LOW EARTH ORBIT 19
   3.2 MARS ENTRY 21
   3.3 EARTH RETURN FROM BEYOND LEO 22
   3.4 EXPLORATION OF VENUS AND OUTER PLANETS 24
   3.5 EXPLORATION OF TITAN 25
   3.6 MATERIAL TECHNOLOGY ADVANCEMENT 26

4.0 CURRENT CAPABILITIES AND LIMITATIONS 28
   4.1 BASIC ARC JET OPERATION 29
   4.2 SIMULATION REGIME 30
   4.3 STAGNATION TESTING 32
   4.4 SHEAR TESTING 34
   4.5 PANEL TESTING 37
   4.6 POWER SUPPLY 39
   4.7 SUMMARY 40

5.0 ARC JET CAPACITY 41
   5.1 WORKFORCE 41
   5.2 THROUGHPUT 42
      5.2.1 Ames Research Center 42
      5.2.2 Johnson Space Center 44
      5.2.3 Air Force Arnold Engineering Development Center 44
      5.2.4 Boeing Large Core Arc Tunnel 45
      5.2.5 Testing Costs 45
      5.2.6 Lunar Environment Arc Jet Facility 45
   5.3 INFRASTRUCTURE CONDITION 46
      5.3.1 Ames Research Center 46
      5.3.2 Johnson Space Center 48
      5.3.3 Air Force Arnold Engineering Development Center 50
   5.4 MOTHBALLING THE EXISTING ARC JET COMPLEXES AT ARC AND JSC 51
   5.5 THROUGHPUT EFFICIENCIES 52
   5.6 SUMMARY 52

6.0 CONCLUSIONS AND PLAN FORWARD FOR NASA INVESTMENT IN ARC JET FACILITIES 54
   6.1 REQUIRED CAPABILITY 54
   6.2 OPTIONS FOR NEW CAPABILITY 57

6.3	Considerations for Upgrading to Future Capability	60
6.4	Technology Development Requirements	61
6.5	Certification and Mission Assurance Roadmap	63
6.6	Transition and Institutional Management	63
6.7	Development Schedule	64
6.8	Summary	65

7.0 FINDINGS AND CONCLUSIONS 67

APPENDIX A: ACRONYMS 70

APPENDIX B: GLOSSARY 73

APPENDIX C: REFERENCES 74

APPENDIX D: STUDY INFORMATION 75

APPENDIX E: AJEWG TEAM BIOS 78

APPENDIX F: CURRENT ARC JET CAPABILITIES 85
   AMES RESEARCH CENTER ARC JET COMPLEX 85
      Interaction Heating Facility (IHF) 86
      Aerodynamic Heating Facility (AHF) 90
      Panel Test Facility (PTF) 93
      Turbulent Flow Duct (TFD) 95
   JOHNSON SPACE CENTER ARC JET COMPLEX 96
      Test Position-1 (TP-1) 98
      Test Position-2 (TP-2) 100
   ARNOLD ENGINEERING DEVELOPMENT CENTER ARC JET COMPLEX 100
   BOEING LCAT ARC JET COMPLEX 103

APPENDIX G: INVESTMENT OPTIONS 106
   1.1 CLOSE THE EXISTING ARC JET COMPLEXES AT JSC AND ARC 106
   1.2 CONTINUE OPERATING BOTH ARC JET COMPLEXES BUT PROVIDE NO ADDITIONAL INVESTMENT FOR REVITALIZATION AND RECAPITALIZATION 107
   1.3 MAKE BOTH ARC AND JSC COMPLEXES HEALTHY, INCLUDING INVESTMENT FOR REVITALIZATION AND RECAPITALIZATION 107
   1.4 CLOSE THE ARC COMPLEX AND MAKE THE JSC COMPLEX HEALTHY 108
   1.5 CLOSE THE JSC COMPLEX AND MAKE THE ARC COMPLEX HEALTHY 108

APPENDIX H: WORK STATEMENT 110

List of Figures

FIGURE 2.1. PERFORMANCE COMPARISON OF VARIOUS GROUND-BASED SIMULATION FACILITIES 13
FIGURE 2.2. DESIGN MARGIN POLICY FOR PROTECTING AGAINST BOND-LINE OVER TEMPERATURE. 14
FIGURE 2.3. CFD WAS USED TO IDENTIFY SOURCE OF INTERFERENCE AND TO IDENTIFY A SHAPE CHANGE THAT ADDRESSES THE PROBLEM 17
FIGURE 3.1. EXISTING ARC JETS PROVIDE ADEQUATE CAPABILITY FOR LEO RETURN VEHICLES 20
FIGURE 3.2. HEAT FLUX AND SHEAR LEVELS FOR MDR, LDR AND LEO RETURN TRAJECTORIES 23
FIGURE 3.3. ENTHALPY LEVELS FOR MDR TRAJECTORY ARE DOUBLE THE CAPABILITY OF EXISTING FACILITIES 23
FIGURE 3.4. PROPOSED TESTING REQUIREMENTS FOR QUALIFICATION OF AN ABLATIVE TPS 26
FIGURE 4.1. ARC JET SCHEMATIC 29
FIGURE 4.2. ARC JET FACILITY COMPONENTS 30
FIGURE 4.3. COMPARISON OF NASA, AEDC, AND BOEING ARC JET FACILITY PERFORMANCE WITH SEVERAL NOMINAL EARTH RETURN TRAJECTORIES. 31
FIGURE 4.4. STAGNATION POINT HEAT FLUX VS. STAGNATION PRESSURE FOR A 10-CM HEMISPHERE 33

FIGURE 4.5. STAGNATION FACILITY COMPARISON, CENTERLINE ENTHALPY VS. WALL PRESSURE FOR 10-CM HEMISPHERE. ...34
FIGURE 4.6. SHEAR TEST FACILITY COMPARISON, COLD WALL HEAT FLUX VS. WALL PRESSURE ...35
FIGURE 4.7. SHEAR TEST FACILITY COMPARISON, COLD WALL HEAT FLUX VS. SHEAR FORCE. ...36
FIGURE 4.8. SHEAR TEST FACILITY CAPABILITY COMPARED WITH SEVERAL BALLISTIC TRAJECTORIES (TRAJECTORIES SHOWN WITH FULL MARGINS). ...37
FIGURE 4.9. PANEL TEST CAPABILITY SHOWING HEAT FLUX VS. PRESSURE. ...38
FIGURE 4.10. PANEL TEST CAPABILITY SHOWING HEAT FLUX VS. PRESSURE WITH ADDITION OF THE TFD. ...39
FIGURE 5.1. ARC ARC JET COMPLEX CUSTOMER BASE ...43
FIGURE 5.2. AEDC PROPOSED UPGRADE TO MID-PRESSURE ARC HEATER ...51
FIGURE 6.1. ESTIMATED 30-YEAR COST OF OPTIONS ...60
FIGURE F-1. AERIAL VIEW OF AMES ARC JET COMPLEX. ...85
FIGURE F-2. AMES INTERACTION HEATING FACILITY (IHF). ...87
FIGURE F-3. 10-DEGREE CONICAL NOZZLE SEGMENTS FOR IHF. ...87
FIGURE F-4. MODEL INJECTION SYSTEM SHOWING TWO STING ARMS. ...87
FIGURE F-5. IHF STAGNATION PERFORMANCE FOR COLD WALL HEAT FLUX VS. PRESSURE. ...88
FIGURE F-6. IHF STAGNATION PERFORMANCE FOR CENTERLINE ENTHALPY VS. PRESSURE. ...88
FIGURE F-7. IHF WEDGE TEST CAPABILITY ON A 4-IN X 4-IN SAMPLE, HEAT FLUX VS. WALL PRESSURE. ...89
FIGURE F-8. IHF WEDGE TEST CAPABILITY ON A 4-IN X 4-IN SAMPLE, HEAT FLUX VS. SHEAR FORCE. ...89
FIGURE F-9. IHF PANEL TEST PERFORMANCE ON A 24-IN X 24-IN SAMPLE, HEAT FLUX VS. WALL PRESSURE ...90
FIGURE F-10. AMES AERODYNAMIC HEATING FACILITY (AHF). ...91
FIGURE F-11. AHF MODEL INJECTION SYSTEM. ...91
FIGURE F-12. AHF STAGNATION PERFORMANCE, HEAT FLUX VS. PRESSURE. ...92
FIGURE F-13. AHF STAGNATION PERFORMANCE, CENTERLINE ENTHALPY VS. PRESSURE. ...92
FIGURE F-14. PANEL TEST FACILITY TEST CABIN SHOWING SEMI-ELLIPTIC NOZZLE AND TEST PANEL. ...93
FIGURE F-15. PTF PANEL TEST PERFORMANCE ON A 24-IN X 24-IN SAMPLE, HEAT FLUX VS. WALL PRESSURE. ...94
FIGURE F-16. TPTF PANEL TEST PERFORMANCE ON A 4-IN X 4-IN SAMPLE, HEAT FLUX VS. WALL PRESSURE. ...94
FIGURE F-17. TPTF TEST PERFORMANCE ON A 4-IN X 4-IN SAMPLE, HEAT FLUX VS. SHEAR FORCE. ...95
FIGURE F-18. TURBULENT FLOW DUCT. ...96
FIGURE F-19. TFD PERFORMANCE ON A 2-IN X 9-IN SAMPLE, HEAT FLUX VS. WALL PRESSURE. ...96
FIGURE F-20. SKETCH OF THE JSC ARC JET COMPLEX. ...97
FIGURE F-21. JSC ARC JET FACILITY SHOWING TP-1 ON THE LEFT AND TP-2 ON THE RIGHT. ...98
FIGURE F-22. TP-1 CHANNEL NOZZLE SHOWING THREE TEST LOCATIONS. ...99
FIGURE F-23. TP-1 PANEL TEST PERFORMANCE FOR HEAT FLUX VS. WALL PRESSURE. ...99
FIGURE F-24. TP-2 40-IN. EXIT DIAMETER NOZZLE CONFIGURATION. ...100
FIGURE F-25. AEDC H2 HUELS TYPE ARC HEATER. ...101
FIGURE F-26. VIEW OF AEDC H2 FACILITY TEST CABIN AND MODEL INJECTION SYSTEM. ...101
FIGURE F-27. H2 WEDGE TEST PERFORMANCE COMPARED WITH NASA WEDGE TEST PERFORMANCE. ...102
FIGURE F-28. H2 STAGNATION TEST PERFORMANCE COMPARED WITH NASA FACILITY PERFORMANCE. ...102
FIGURE F-29. VIEW OF BOEING LCAT FACILITY. ...103
FIGURE F-30. VIEW OF LCAT TEST CABIN INTERIOR AND MODEL INJECTION SYSTEM. ...103
FIGURE F-31. LCAT STAGNATION PERFORMANCE OF HEAT FLUX VS. PRESSURE. ...104
FIGURE F-32. LCAT STAGNATION PERFORMANCE OF CENTERLINE ENTHALPY VS. PRESSURE. ...104
FIGURE F-33. LCAT WEDGE TEST PERFORMANCE, HEAT FLUX VS. WALL PRESSURE. ...105
FIGURE F-34. LCAT WEDGE TEST PERFORMANCE, HEAT FLUX VS. SHEAR FORCE. ...105

List of Tables

Table 3.1. Parameters. ...19
Table 3.2. Representative Earth Entry Conditions for Various Missions. ...24
Table 5.1. Throughput for ARC Arc Jet Complex. ...43
Table 5.2. Throughput for JSC Arc Jet Complex. ...44
Table 5.3. The Cost of Arc Jet Testing. ...45

Table 5.4. ARC Arc Jet Complex Subsystem Summary..47
Table 5.5. JSC Arc Jet Complex Subsystem Summary..49
Table 5.6. JSC Arc Jet Complex System Upgrades...49
Table 6.1 Future Arc Jet Needs: Summary of Responses from End Users......................................55
Table 6.2. Regional Construction Cost Index...59
Table 6.3. Estimated 30-Year Costs for a New Arc Jet Complex..60
Table 6.4. Notional Schedule for Transition to New Arc Jet Capability..66
Table F-1. ARC Arc Jet Complex Active Configuration Summary..87
Table G-1. Comparing Costs of Various Options..111
Table H-1: Program and Project Requirements for Arc Jet Capability...114

1.0 Introduction

The Arc Jet Evaluation Working Group (AJEWG) was chartered to provide engineering support to the NASA Chief Engineer, Mike Ryschkewitsch, in evaluating the availability and use of hypersonic thermal arc jet test facilities to meet NASA's planned and future mission requirements over the next 30 years. NASA has a sustained mission to develop and certify entry systems technology for these mission categories:

- LEO return
- Mars entry
- Earth return from beyond LEO
- Exploration of Venus and outer planets
- Titan

In addition, the Nation has a sustained interest in long-duration hypersonic flight and air-breathing space access, including the activities of the National Partnership for Aeronautical Testing (NPAT) and the Hypersonic Propulsion Integrated Testing Team (HPITT). Congress mandated that DOD and NASA work together on hypersonics with the Joint Technology Office on Hypersonics (JTOH).[1]

The AJEWG was to provide findings to inform possible long-term investment strategy alternatives for NASA's arc jet capability.

1.1 Scope

The AJEWG was asked to consider both NASA and non-NASA arc jet hypersonic thermal test facilities to meet NASA mission requirements. The AJEWG looked at existing facilities and evaluated the capabilities of facilities capable of supporting large-scale thermal protection system (TPS) development and certification. Smaller research and development arc jets and arc jets used for materials screening at the NASA Langley Research Center (LaRC), the Arc Heated Scramjet Test Facility (AHSTF), and the Hypersonic Material Environmental Test System (HyMETS) were not considered viable possibilities. The AJEWG considered it unrealistic to rely on foreign facilities like Scirocco in Italy to support NASA's leadership in development and modeling of TPS materials. Aside from issues of logistics and control of schedule, restrictions imposed by the International Traffic in Arms Regulations (ITAR) and security considerations make use of the complex untenable.

1.2 Background

Arc jets facilities are designed to simulate conditions for space vehicles entering through a planetary atmosphere. They provide ground-based testing for re-entry vehicle thermal protection materials, by electrically heating air or other test gases to a flight-like enthalpy. This high-energy

---

[1] The John Warner National Defense Authorization Act for Fiscal Year 2007, P.L. 109-364, Section 218, directed the Secretary of Defense to establish within the Office of the Secretary of Defense the JTOH. The JTOH is to coordinate with the programs on hypersonics of the National Aeronautics and Space Administration and is required to report to the Congressional Defense Committees a roadmap for the hypersonics programs of the Department of Defense every two years. The first roadmap came out in February 2008.

gas is expanded through a nozzle at high velocity into a vacuum test chamber where a test article or articles are exposed to the heat rate, pressure, and shear that simulate the conditions of a given flight trajectory.

The Ames Research Center (ARC) and the Johnson Space Center (JSC) facilities were brought online in the mid-1960s for development of thermal protection material and mission support. The JSC facilities were funded through the Space Transportation System (STS) Program (U.S. Space Shuttle program) until 2002, but the amount of sustaining engineering testing at that time was significantly decreased, and the JSC arc jets were designated to be mothballed. Following the Columbia accident, both JSC and ARC facilities were heavily utilized in the investigation and in the subsequent Return to Flight (RTF) activities. The Columbia Accident Investigation Board (CAIB) recommendations included development of a TPS repair capability. The arc jets played a major role in demonstrating the effectiveness of proposed repairs, including tests of "the full-scale repair components at maximum heating for the full duration of the flight trajectory." [ARC, Basis of Need] (References are listed in Appendix C: References.)

1.3 Evaluation Approach

The Office of the Chief Engineer (OCE) asked the AJEWG to identify and document requirements for NASA missions and aeronautics research that would necessitate the use of arc jet facilities that perform hypersonic thermal testing and to evaluate these requirements against the performance capability and throughput capacities of existing facilities.

To accomplish the objective, the AJEWG conducted site visits at both JSC and ARC and listened to presentations on the facilities and from users. (See Appendix D: Study Information.) The working group also reviewed previous arc jet evaluations, relevant reference documents, current maintenance plans, and written justifications and plans for the proposed upgrade to the Lunar Environment Arc Jet Facility (LEAF).

During the site visits, interviews were also conducted with key technical leads working all stages of the TPS life cycle, from entry system precursor technologies, through flight vehicle development, to sustaining engineering of operational vehicles. On site, the working group observed and assessed existing arc jet complex support infrastructure, facility hardware, test diagnostics, workforce, operational efficiency, maintenance schedules, and upgrade requirements.

Arc jet operations at the Arnold Engineering Development Center (AEDC) and Boeing Large Core Arc Tunnel (LCAT) were reviewed by interviewing key operators from both complexes. A visual comparison of the capabilities and trajectories covered by these different facilities is made in Section 4.0, Current Capabilities.

The AJEWG also assessed requirements for future arc jet testing capability and capacity by interviewing current and potential users of arc jet services from the NASA Exploration Systems, Space Operations, Science, and Aeronautics Mission Directorates; the Air Force; and SpaceX and commercial organizations contracted with NASA for entry systems.

The amount of testing accomplished, or throughput; planned maintenance; and facility upgrades were considered. The findings and conclusions of this analysis are captured in Section 5.0 of this report. The way forward for a possible investment strategy for NASA's hypersonic thermal test facility capability is considered in Section 6.0.

The AJEWG members were selected from among national experts in arc jet technology and entry technology to provide an unbiased perspective and a focused, independent evaluation to inform NASA's long-term investment strategy toward arc jet capability. (See Appendix E: Team Bios.)

1.4 Current Changing Climate for Funding and Requirements

The assignment of this evaluation report occurred at a time of flux in Agency mission requirements and funding. The context for this report includes the recent establishment of the Facilities Program Board (FPB), whose charter is to provide a comprehensive Agency facility strategic direction. ("Facilities" is used here to indicate Agency buildings, assets, and other infrastructure. The arc jet community uses "complex" and "facility" interchangeably, but this document attempts to consistently use "complex" to refer to the arc jets and their supportive infrastructure and "facility" to refer to an arc jet tunnel.) The FPB established an Agency facility strategy to reduce the proportion of facilities within the Agency that are older than 40 years and beyond their design life. Despite such program planning challenges as the Constellation redirection and retirement of the U.S. Space Shuttle, the AJEWG concluded that NASA's arc jet capability serves a persistent strategic need for the Agency.

1.5 Document Organization

This report lays out the engineering uses of arc jets in Section 2.0. Section 3.0 describes the requirements of different mission categories. The current state of the main national arc jet complexes is described in Section 4.0, along with the gaps in arc jet technology that should be addressed.

The analysis of the throughput and a comparison of the workforce at different arc jet complexes is captured in Section 5.0. Section 6.0 explores investment options for the Agency, focusing on future capability and capacity that would meet the needs for a safe return from Mars. Findings and conclusions appear at the end of each section and are collected in the final Section 7.0 of the report.

2.0 Engineering Uses and Accomplishments of Arc Jets

Spacecraft entering into a planet's atmosphere at high velocity are subjected to extreme aerothermal loads. Protecting against these extreme loads is a critical element of spacecraft design. With the possible exception of very limited locations on Shuttle, entry vehicle thermal protection failure is single point and catastrophic to mission success. Human-rating requirements for space systems generally specify that systems should be fault tolerant to catastrophic events [NPR 8705]. Fault tolerance design avoids failure through redundancy or other compensation. However, vehicle TPS is often exempted from this requirement because redundancy is not possible. The risk is instead controlled through defined standards for increased design margin. For robotic missions, mission assurance requirements will drive similar approaches to TPS for increased margin. The environments that arc jets can simulate allow material engineers and spacecraft designers to:

- Design, develop, qualify, and certify spacecraft thermal protection systems (TPS), including the heat shield, backshell, and seal design;
- Screen and develop new thermal protection materials;
- Certify vehicles for flight and establish performance margins for the TPS design during entry, descent, and landing (EDL);
- Re-qualify materials that have new manufacturing facilities, processes, vendors, or constituent materials;
- Perform sustaining engineering throughout the TPS mission life, including advancing and improving the TPS;
- Test, analyze, and perform trades to support decisions on safe return or repair alternatives to on-orbit anomalies in the TPS; and
- Support the investigation of TPS failures or mishaps.

For more than 40 years, arc jet testing has been the primary basis for characterizing TPS in support of material development and response model validation. Every NASA atmospheric entry mission has relied on arc jet testing for TPS development. Arc jet facilities provide the only ground-based means of simulating entry heating rates in a reacting flow environment for flight-relevant durations, as shown in the Figure 2.1 below. Although shock tunnels can more accurately simulate the aerothermal environment, they do so for too short a period to permit accurate assessment of the material response to that environment. Exposure durations approximating those expected during flight are required to screen and qualify thermal protection materials. Screening typically subjects material samples to flight-relevant heat fluxes and pressures, but rarely addresses all aspects of the load environment that might contribute to material failure. In the Apollo era, for example, when arc jet power levels did not exceed 20 MW, stagnation testing of small models was performed, but shear testing of large models was not attempted.

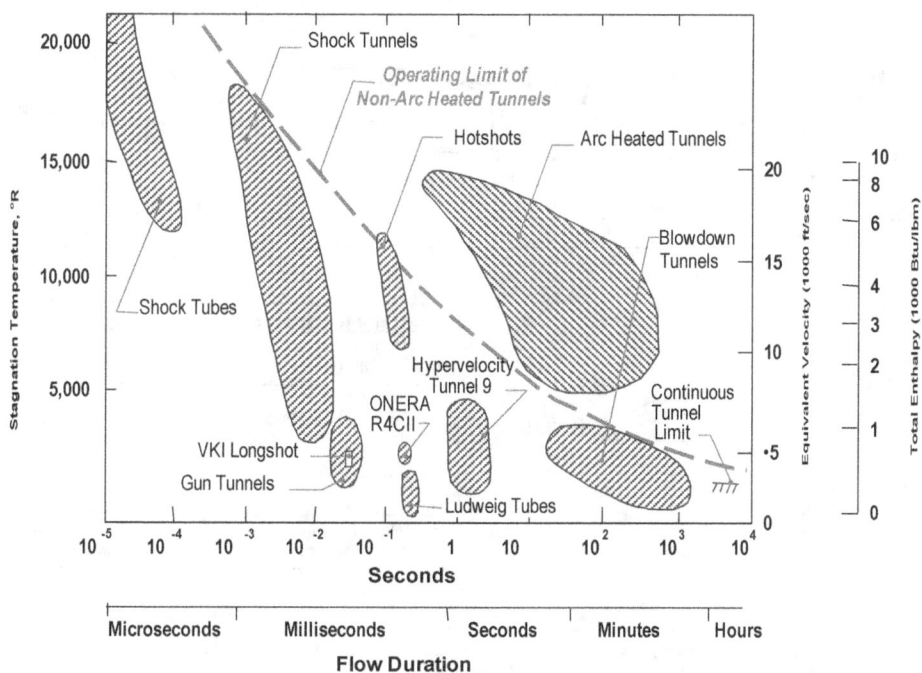

Figure 2.1. Performance Comparison of Various Ground-based Simulation Facilities [Curry, Past Experience Using Arc Jet Testing for Human Spacecraft TPS]

Thermal protection failure modes are typically of two types: bond-line over temperature due to under-predicted heat loads or over-predicted material response, or catastrophic overheating due to mechanical failure [Steltzner, Mars Exploration Program Use Profile]. Bond-line temperature can generally be margined by increasing material thickness, at the expense of increased mass. Arc jet testing provides data for detailed material response models that can reduce uncertainty and the magnitude of thickness margins. Arc jets can also uncover mechanical failure modes including erosion, spallation, and loss of liquid layer due to shear or failure around a heat shield penetration. Heat shield penetration failures are difficult to identify, characterize, and control, but arc jet testing can help to expose subsystem feature failure modes early in the design.

In the recent Constellation TPS Advanced Development Project, a formal margins policy, which is summarized in Figure 2.2 below, was generated to support estimation of subsystem reliability, but it addresses only a bond-line over temperature failure mode. Arc jet testing was also conducted to provide qualitative information on other failure modes, including local burn through from damage, flow within a porous low-density ablator material, and flow ingestion due to seal failure.

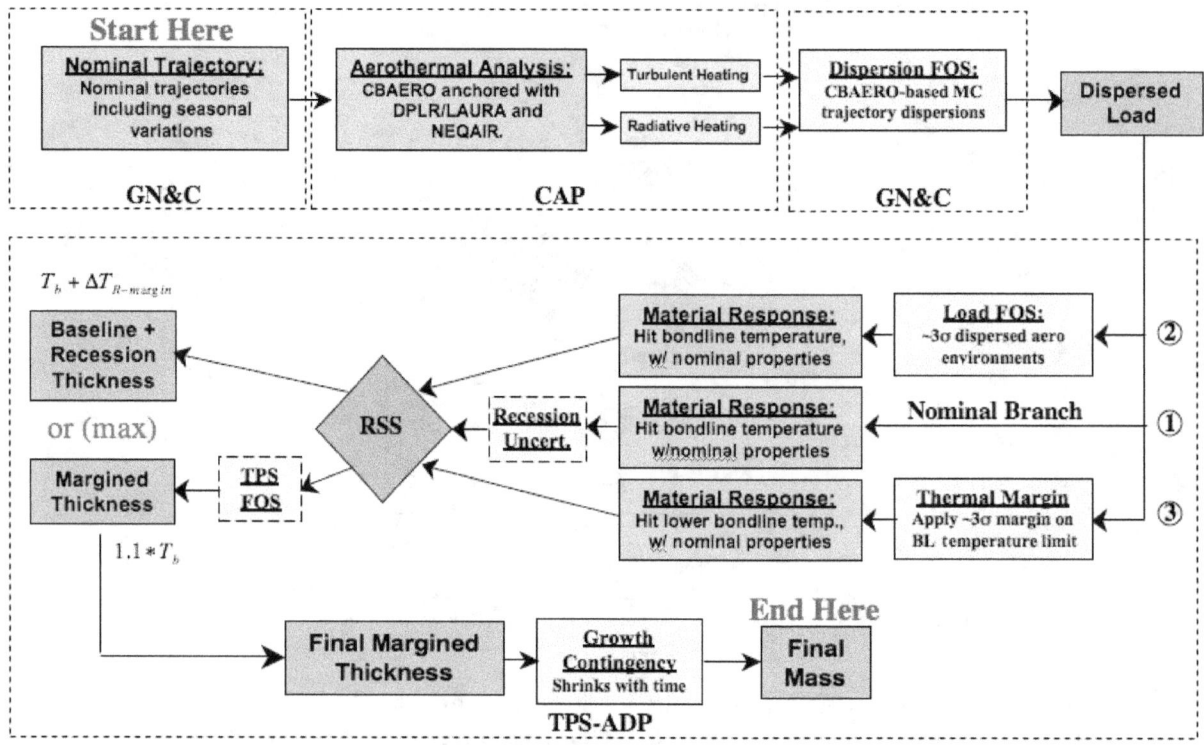

Figure 2.2. Design margin policy for protecting against bond-line over temperature. [Wright, Anticipated Arc Jet Usage]

2.1 Material Response Modeling

The design and qualification of thermal protection materials for entry vehicles and probes requires consideration of complex thermal and aerodynamic loads, which are strongly affected by the chemical state of the atmospheric gases and their interaction with the vehicle surface. The primary driving need for NASA to have arc jet test capability is to support the development of thermal protection materials, and the qualification of those materials for flight, by providing the best possible ground test simulation of the flight environment. Extensive arc jet tests have supported the development of all thermal protection materials used by NASA. Concurrent with material development, numerous arc jet tests were needed to anchor response models, which are used to assess material behavior and establish design margins. The development of thermal protection materials and response models will be persistent strategic needs for NASA well into the future. Arc jets provide essential ground test support for these needs.

The selection of appropriate TPS material is driven by mission environment and thermal management needs of a vehicle. Material selection is typically governed by entry conditions for peak heat flux, stagnation pressure, and shear force; whereas material thickness is dictated by the integrated heat load and bond-line temperature capability. For modeling purposes, the total calculated heat flux must include augmented heating from surface catalytic reactions, roughness effects, turbulent flow, and shock layer radiation. Thermal protection materials basically fall into either a single-use or multi-use category. Single-use materials are typically ablators, which accommodate high-entry heat load conditions, 100 to 30,000 W/cm^2, through phase change, mass loss, and char radiation. Multi-use materials are typically refractory materials that manage more moderate heating conditions, 10 to 100 W/cm^2, by radiating heat back into a flow field with limited chemical reaction or mass loss.

For single-use materials, modeling ablation is extremely complex because the process often involves pyrolysis of organic constituents, in-depth production of volatile gases, and oxidation-reduction of carbon and silicon. Therefore, ablation behavior is strongly affected by material constituents, in-depth microstructure, and surface interaction with the entry flow [Laub et al., Past Experience: A Different Perspective]. The complexities of phase change at the pyrolysis zone, diffusion of gases through a char layer, and gas ejection into the flow boundary layer present significant challenges to state-of-the-art TPS design methodology. Current design models that are used to characterize ablators for heat shields rely on phenomenological calibration with a database of arc jet test results. Although no ground-based thermal test can simulate actual flight conditions, the high-temperature reacting flow within an arc jet is recognized as the best capability for characterizing ablator response and calibrating design models.

A current limitation with ablation modeling is the often important design requirement to characterize failure thresholds. Unfortunately, the rigorous experiments needed to define those thresholds and identify critical failure mechanisms are rarely performed. As a result, defining design "margin" is quite arbitrary since the conditions leading to failure are not known [Venkatapathy et al., Capabilities Needed]. Improved understanding of the physical processes underlying ablation will enable a greater understanding of the margin between safe operation and failure. More accurate models combined with a fundamental understanding of ablator failure physics may be the only means of reducing heat shield mass without sacrificing operational reliability. Accomplishing this end will require carefully controlled arc jet experiments.

In contrast, with ablators, the failure thresholds for multi-use materials are typically well known, due in large part to the fact that heating loads are moderate and material property changes are minor after successive mission exposures. Although far simpler than ablator materials, arc jet testing is still critical for characterizing multi-use materials and calibrating design models. Multiple exposures are needed to support life cycle analyses, which rely on known and well-characterized failure thresholds. Multi-use materials are tested to establish a failure threshold and the heat shield is then designed to operate with a safe margin from that threshold. This design approach played a strong role in the CAIB failure reconstruction effort and the subsequent RTF activities. Based on CAIB recommendations, extensive arc jet tests were performed to assess the thermal protection system's ability to sustain flight debris damage and safely return to Earth.

2.2 Computational Fluid Dynamics Modeling

Computational Fluid Dynamics (CFD) is used routinely in TPS design. Reacting flow codes, such as Data-Parallel Line Relaxation (DPLR) and Laura, are used to predict environments around the vehicle throughout entry. The predicted thermal, pressure, and shear loads are input to material response codes, such as Fan/Inlet Acoustic Technology (FIAT), Cryogenic Moisture Apparatus (CMA), and Structural and Thermal Analysis Branch (STAB), to calculate the thickness of material required to keep the underlying vehicle structure at acceptable temperatures. These same CFD codes can be applied to simulate the flow in an arc jet facility. Good correlation between simulated and measured conditions in the arc jets generates confidence in the predictive power of the CFD codes and strengthens the traceability for flight simulation. Furthermore, success in modeling the flow in these facilities may extend their role beyond data generation for material characterization and into aerothermodynamic flow testing.

Since CFD does not address material response (which is modeled separately), it is unrealistic to imagine that CFD can replace ground testing for TPS design. Instead, CFD can be used in two ways to improve the material response information that is generated in arc jet tests:

- Reduce uncertainty in the environment to which the model is exposed;
- Predict conditions for a range of arc jet power and mass flow settings, to assist in designing a test series that subjects the material to conditions that exercise all failure modes

The current state of the art in facility simulation does not model the arc heater directly, but assumes that it provides high-enthalpy gas into a plenum, and this slug of gas is expanded through a convergent-divergent nozzle into the test section. This way, the computational process is no different than one would employ for hypersonic wind tunnels, although the thermochemical modeling of the gas is considerably more complex. Arc-heated flow fields are computed and calibrated against calorimetric and pitot measurements, and optical measurements (when available). Apart from providing details of the flow fields and flow structures, the important results of the simulations are the nozzle centerline total enthalpy (a key parameter used in computing a material's thermal response), shear stress, and hot-wall heat flux, all three of which are not measured directly in arc jet tests. By reducing the uncertainty in these parameters, the CFD supports more precise modeling of material response sensitivity to these quantities.

The contributions of CFD can be enhanced through greater use of flow diagnostics in the arc jet facilities, both in the plenum and the test section. Techniques such as optical emission spectroscopy and laser-induced fluorescence (LIF) provide knowledge of the thermochemical state of the gas/gas mixture and produce valuable information on flow non-uniformities and asymmetries that are currently not modeled. Measurement of thermal losses in the core will characterize the validity of the current assumption that the core is adiabatic: if measurements indicate significant losses, the assumption will need to be modified.

Apart from characterizing the flow in a particular test, CFD can be used to design models and test conditions that produce desired heating, pressure, and shear at the model surface. An early example of successful three-dimensional flow modeling [Prabhu, Modeling as an Alternative] is illustrated in Figure 2.3 below. A nose tip model at angle of attack had serious flow interference at the base of the model. Through a series of simulations for a range of sting and model shapes, CFD provided insight into a geometry change that fixed the interference issue.

It is anticipated that CFD will be used more routinely in the future to design models that maximize the flight relevance of conditions produced in an arc jet facility of a given capability.

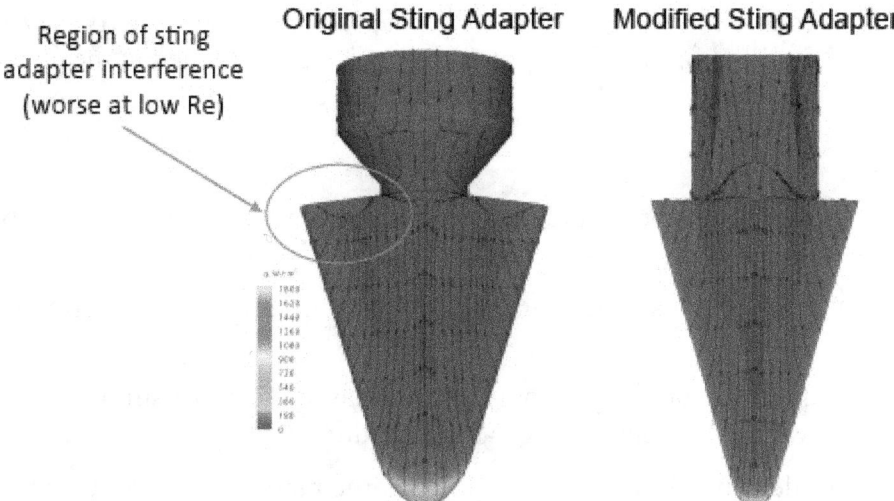

Figure 2.3. CFD was used to identify source of interference and to identify a shape change that addresses the problem [Prabhu, Modeling as an Alternative]

2.3 Maximizing Value of Arc Jet Testing

TPS design is typically regarded as trading subsystem reliability against mass. This is a reasonable perspective when failure mechanisms are well understood, and the required margin is driven by the level of uncertainty, both in the applied load and the material response. Instrumentation and flow diagnostic schemes that measure in-situ arc jet conditions can drive down uncertainty in the applied load. Techniques that provide a more detailed view of material response, including species identification downstream of the test article, video of the model surface that indicates time-accurate ablation behavior, and interior instrumentation of the test article can provide physical insight to drive improvements in the material and the response model.

Several flow diagnostic tools have been available for two decades, and low-level research activities have been conducted for much of that period, but the diagnostics have not been integrated into standard arc jet test practices. This situation appears to be largely driven by the lack of access that researchers have had to the arc jet facilities, due to a lack of funding, and to higher prioritization being given to tests that directly support mission development and operations. Diagnostics have not been sufficiently mature to be adopted by individual missions. Strategic management of arc jet facilities as agency assets can reprioritize instrumentation development to increase the value of arc jet testing for all missions.

Where failure mechanisms are not fully characterized, additional mass may not improve reliability. A failure that quickly removes material after it is initiated, such as shear-induced erosion, or one which bypasses the bulk material, such as failure of gap fillers, must be addressed by altering the design concept rather than by simply increasing thickness. A comprehensive margin policy must address all failure modes, which requires testing that establishes performance bounds for all modes. Such testing has not been routinely conducted by missions, which are primarily concerned with performance at, or slightly above, the conditions predicted for their specific mission trajectories. The scope of screening tests can be expanded and the level of instrumentation can be increased, to identify and understand failure mechanisms. Comprehensive failure investigation can be conducted as part of a material development activity. As with instrumentation and facility improvements, such testing is best managed at the agency level, to establish a suite of fully characterized materials that are available to all missions.

2.4 Summary

Finding 2.1: Arc jet testing has proven to be a core competency and required capability of NASA.

Finding 2.2: Every NASA atmospheric entry mission has relied on arc jet testing for TPS development.

Finding 2.3: Arc jet facilities provide the only ground-based means of simulating entry heating rates in a reacting flow environment for flight-relevant durations.

Finding 2.4: Ablator design policy currently applies large margins to cover uncertainty in applied loads and material response. Improvements in arc jet diagnostics and test article instrumentation can reduce uncertainties and hence reduce margins.

Conclusion 2.1: Existing ablator design policy is incomplete, because some failure modes are not fully understood, so their contribution to system reliability is not quantified. More rigorous examination of material failure modes will enable more defensible mission assurance assessment. Characterization of failure modes is better managed as a material development activity rather than being conducted for individual missions.

Conclusion 2.2: Currently available diagnostic capabilities and instrumentation techniques should be infused into NASA arc jet standard practice. This infusion is best managed as an Agency strategic investment rather than as a programmatic responsibility.

3.0 Mission/Future Requirements/National Needs

The thermal protection needs for different mission classes are strongly affected by the entry velocities and atmospheric properties that prevail for each class. The mission parameters that drive arc jet capability are summarized in Table 3.1 below and discussed in more detail in the following subsections.

The acceptable level of risk and the certification strategy for each mission type will drive the amount of testing that is required, and hence, the arc jet capacity requirements for the Agency.

Table 3.1. Parameters
**[Constructed with data from Munk, Future Missions;
Venkatapathy et al., Outer Planet Missions]**

Characteristics	LEO Return	Mars entry	Earth return from beyond LEO (comet)	Exploration of gas giants (Saturn)	Titan
Pressure	85 kPa	25 kPa	170 kPa	300 kPa	10 kPa
Heat Flux					
Convective	400 W/cm^2	500 W/cm^2	3000 W/cm^2	4500 W/cm^2	60 W/cm^2
Radiative	N/A	100 W/cm^2	7500 W/cm^2	100 W/cm^2	40 W/cm^2
Shear	200 Pa	250 Pa	500 Pa		100 Pa
Flow State					
Turbulent	Nice to have	Yes	Yes	Nice to have	Nice to have
Laminar	Yes	Yes	Yes	Yes	Yes
Enthalpy	25MJ/kg	30MJ/kg	100 MJ/kg		15MJ/kg
Gas Constituents	Air	CO_2	Air	H2/He	N_2/CH_4
Model Size	50 cm	Large enough to test system features and closeouts	Large enough to test system features and closeouts	10 cm	10 cm

3.1 Return from Low Earth Orbit

Capabilities of existing NASA arc jet facilities are well matched to LEO return mission requirements. Figure 3.1 shows that typical trajectories for lifting bodies entering the atmosphere at about 7 km/s lie well within the range of conditions that can be simulated in these facilities. Existing facilities have also been used throughout the life cycle of crew transportation vehicles, and are still used for real-time support of each Shuttle mission (testing was conducted in support of STS 117 and 118 to guide decision makers on the need for TPS repair prior to re-entry).

The amount of testing that will be required for LEO return is highly uncertain, because human rating requirements (HRR) for commercial crew transport and cargo are not well-defined. The 2009 report from the NASA Aerospace Safety Advisory Panel (ASAP) states:

> "It is the Panel's position that no COTS manufacturer is currently HRR qualified, despite some claims and beliefs to the contrary. Questions that must be answered

are: What is the process for certifying that potential COTS vehicles are airworthy and capable of carrying astronauts into space safely? How is compliance assured over the life of the activity? The same questions would apply to any potential international orbital transportation systems."

In response to the Panel's report, NASA agreed "to perform additional research to support development of the HRR implementation handbook."[2]

If the HRR practice from previous programs is applied for commercial crew vehicles, the amount of testing will be significant and will continue through the operational phase of the life cycle. NASA Administrator Bolden suggests that NASA will "support ... the commercial spaceflight industry to enable hundreds, even thousands of people to visit or live in LEO."[3] Certainly, arc jet test capability will be required for such operations. Certification guidelines must be established before the required test capacity can be accurately predicted.

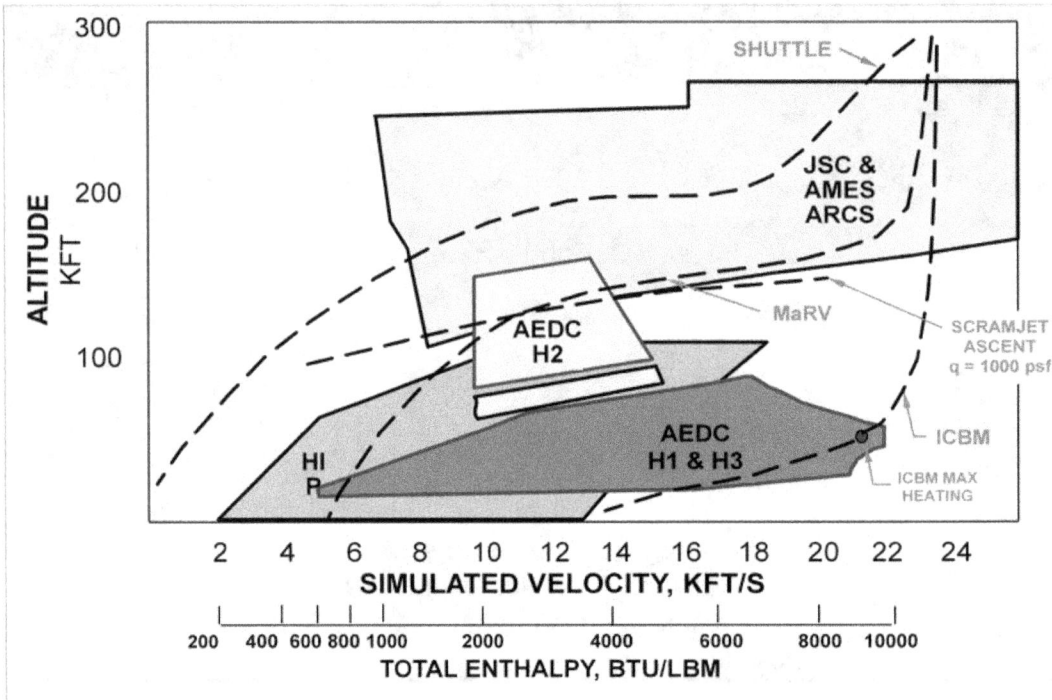

Figure 3.1. Existing arc jets provide adequate capability for LEO return vehicles
[Smith, AEDC Arc Jet Facility Capabilities]

In addition to NASA mission needs, the Air Force Research Laboratory (AFRL) has identified an ongoing need for testing in NASA facilities [Bowman, Arc Jet Testing Perspective]. In recent years, they have had 3 or 4 weeks at ARC for tests of hypersonic glide and cruise vehicles. Although these are not strictly orbital entry vehicles, they operate in an altitude and velocity regime that is better matched to NASA arc jet test capabilities than to the high pressure facilities at AEDC. AFRL anticipates a similar need in the next decade, with the possibility of moderate increases.

---

[2] HRR Handbook described in NASA's response to ASAP recommendation 2008-04-03.
[3] Bolden comments on budget proposal for 2011.

3.2 Mars Entry

Entry velocity at Mars is relatively modest, with a maximum of 7.5 km/s for a direct trajectory. The challenge is to decelerate sufficiently quickly in the thin Martian atmosphere, which is about 1% of the density of Earth's atmosphere, to permit parachute deployment at a safe altitude for terminal descent. In order to generate adequate drag during hypersonic entry, a rigid aeroshell must have a large cone angle which has the adverse effect of increasing the heating rate and turbulent shear stress on the leeward sections of the heat shield.

There have been six successful U.S. landings on Mars, beginning in 1976 with the successful deployment of Viking 1 and 2. The Viking mission, and the EDL technology developed for it, established the backbone for all U.S. Mars missions to date. All U.S. Mars missions have utilized a spherically blunted 70-degree half-angle cone for the forebody and Super Lightweight Ablator (SLA)-561V ablator as the heat shield material. The SLA-561V material has had extensive arc jet testing that, together with the experience gained from successful flights, has qualified it for flight in the thin carbon dioxide atmosphere of Mars.

The Mars Science Laboratory (MSL) mission, currently planned to land in 2012, is the largest and heaviest vehicle yet designed for Mars entry. The mission also has the highest entry velocity of 7.5 km/s, and requires a large-diameter aeroshell for drag. The combination of high-entry velocity, large-diameter aeroshell, and lift ratio is expected to produce conditions where, for the first time, a Mars entry capsule had to be designed for a completely turbulent heat pulse with significant levels of shear stress. Arc jet tests conducted on SLA-561V to assess its capability to sustain conditions predicted for MSL revealed a critical failure mode in which the protective surface layer is swept away by shear loads. Arc jet testing was essential for characterizing this material limitation and additional thickness would not margin against failure. Although SLA-561V had been used successfully on several apparently similar previous missions, its limitation under turbulent shear required a material change from SLA-561 to the Phenolic Impregnated Carbon Ablator (PICA) material for the MSL mission. The change in thermal protection material required significantly more arc jet tests to qualify a new ablator material for Mars, including additional tests to examine the effects of the CO_2 atmosphere on ablation rates

To date, no viable Mars EDL architecture has been put forward that can safely place more than 2 metric tons at the higher surface elevations in close proximity to scientifically interesting terrain. The development of new EDL systems and technologies must begin before human-scale missions to the surface of Mars can be seriously considered. The Entry Descent Landing Systems Analysis (EDL-SA) group recently completed a study to define architecture requirements for delivering 40 metric tons, as will be needed for a crewed Mars mission. The study concluded that both an aerocapture and subsequent entry stage will be required, and that two basic systems, a mid lift-to-drag rigid aeroshell and a Hypersonic Inflatable Atmospheric Decelerator (HIAD), have the potential to land the required mass.

A high lift-to-drag rigid aeroshell vehicle will have a higher ballistic coefficient than the 70-degree sphere cone used to date. At the scale required to deliver large masses to the surface, predicted heating rates are close to 600 W/cm^2, with about 20% of the total heating being radiation from the shock layer. This heat rate is much higher than any previous experience at Mars and will require the development of new thermal protection materials and an aerothermodynamic database, for which a large number of arc jet tests will be required. In addition to the high heat pulse associated with aerocapture, thermal protection materials will also endure an extended cold soak while on orbit before the entry stage is commenced. Thermal protection concepts are proposed to handle the new requirement for dual heat pulse capability, and arc jet tests will be critically important for developing and qualifying those concepts with the required materials for an aerocapture application in the CO_2 atmosphere of Mars.

Deployable decelerators for hypersonic entry are an emerging technology envisioned to provide very large drag area at high altitudes. Just before atmospheric entry, the HIAD is deployed with a flexible TPS cover. The EDL-SA study predicts peak heating rates could be as high as 115 W/cm^2 during the aerocapture stage for a 23-meter diameter HIAD, including a predicted radiative heating level of 40 W/cm^2. The current demonstrated capability for a suitable HAID flexible TPS is less than 30 W/cm^2. Research development efforts funded under the NASA Fundamental Aeronautics Program are pursuing technologies that may extend performance capability to greater than 100 W/cm^2, but arc jet testing support will be a critical assessment tool for capturing the effects of catalysis, chemistry, and flow dynamics. In an effort to reduce uncertainty, thermal performance within a CO_2 environment should be completed to assess the effects of higher oxygen concentrations than Earth.

3.3 Earth Return from Beyond LEO

Entry into Earth's atmosphere from beyond orbit involves velocities of 11 km/s or higher. Conversion of the kinetic energy—which is more than double that for entry from LEO at 7 km/s—to heat involves much higher heating rates. For large, blunt vehicles, the radiative component from the detached shock wave is a significant fraction of the total heating. Figure 3.2 compares the heat flux and shear stress for Lunar Direct Return (LDR) and Mars Direct Return (MDR) with the levels required for LEO return [Reuther presentation of Hash]. Clearly the conditions to be simulated for entry from beyond LEO are far more severe. Also, since a large fraction of the heat flux is radiative, a suitable test facility must produce high levels of both radiative and convective heating to reduce performance uncertainties. Furthermore, radiation absorption by ablation products is an important component of the ablator response so that the radiative heat flux must have representative spectral distributions. Figure 3.3 indicates that enthalpy levels for Mars return are double the levels achievable in existing arc jets. Hence, significant advances in test facility design, beyond the capability of a LEAF-type facility, will be required.

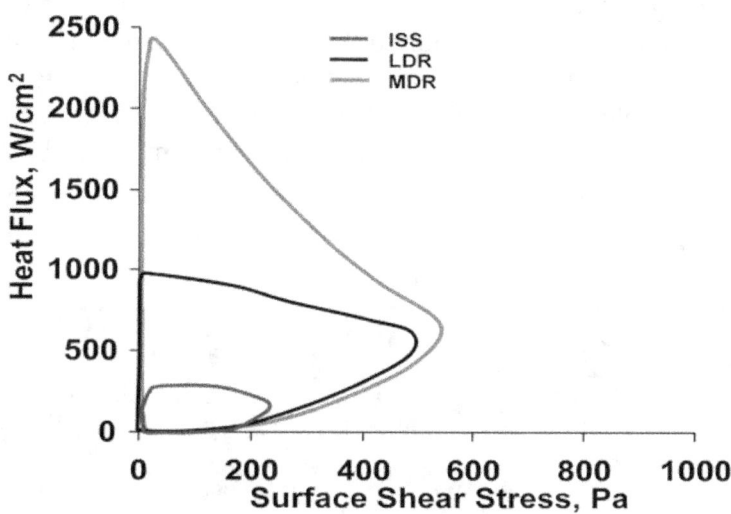

Figure 3.2. Heat flux and shear levels for MDR, LDR and LEO return trajectories
[Hash, Capabilities and Lunar Return Certification Requirements]

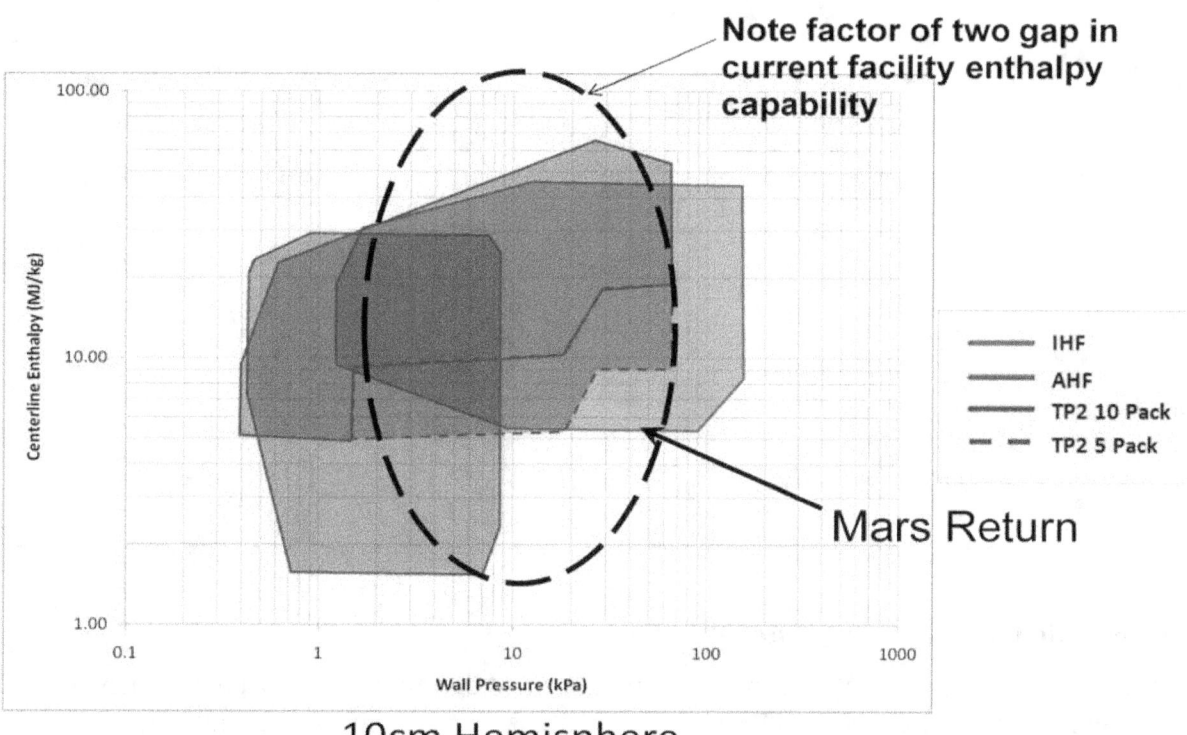

Figure 3.3. Enthalpy levels for MDR trajectory are double the capability of existing facilities
[Wright, Anticipated Arc Jet Usage]

Although the entry conditions for return from beyond LEO exceed the capabilities of existing facilities, it does not follow immediately that such missions cannot be undertaken by NASA without upgraded facilities. The Apollo program used multiple uncrewed flight tests of the full-scale flight vehicle to certify the heat shield for crewed operations. The Orion Project intended a combination of enhanced ground facilities, with the proposed LEAF complex, and flight test to certify the heat shield. The Galileo probe entered the Jovian atmosphere successfully, although the Giant Planet Facility, now defunct, did not generate total heating levels equivalent to those experienced in the mission. Nevertheless, serious consideration of Earth return missions will require significant arc jet capability augmentation, to provide combined convective-radiative heating at elevated heating rates.

Table 3.2. Representative Earth Entry Conditions for Various Missions
[Created using data from Munk, Future Missions]

Mission/Destination	Entry Velocity, km/s	Peak Convective Heating, W/cm2	Peak Radiative Heating, W/cm2	% Radiative Heating of Total Heating	Peak Stagnation Pressure Estimate, Pa	Peak Heating Stagnation Pressure Estimate, Pa
Comet	~14-16	2000-3000	3000-7500	70%	30kPa-300kPa	30kPa-180kPa
Asteroid	~11.5	1250	350	22%	4kPa-150kPa	4kPa-95kPa
Moon	~10-11	700-1000	50-150	15%	4kPa-140kPa	4kPa-85kPa
Mars (Science)	~11-12	1300	600	30%	170 kPa	100kPa
Mars (Human)	~11-12	3000	1400	70%		
Stardust*	12.8	1200	130	10%		27kPa
Genesis*	11.0	700	30	4%		

3.4 Exploration of Venus and Outer Planets

Future exploration missions to Venus and the Outer Planets that require atmospheric entry will place a demanding challenge on the TPS to assure success, with the most demanding requirements being for the Gas Giants. Entry probes circumvent current limitations of remote sensing the inner depths of the atmosphere. To be successful, entry probes must be designed to tolerate the most demanding entry conditions for peak heat rate, total heat load, and dynamic pressure. Enabling future entry probe missions to the Gas Giants will require improvements in both aerothermodynamic modeling and TPS design. Due to extremely high entry velocities, giant planet probes require the most robust thermal protection materials.

The Galileo probe, which was launched with its orbiter in 1989 and entered the Jovian atmosphere in December 1995, was exposed to what is regarded as the most extreme conditions for a TPS material to date. Almost half of the Galileo probe mass of 335 kg was dedicated to the heat shield (152 kg). The probe forebody was a 45-degree angle blunt cone configuration and entered the atmosphere at slightly more than 47 km/sec. In terms of recession tolerance under extreme heating rates and pressure, carbon phenolic, developed by the DOD for ballistic missile entry more than 40 years ago, is still considered to be the most robust ablator today. The combination of entry speed and dense Jovian atmosphere was estimated to yield a combined convective and radiative heating rate as high as 35,000 W/cm^2, and a significant portion of peak heating was due to radiant heating from the bow shock. As was predicted by CFD analyses, heat shield TPS performance flight data relayed from the Galileo probe to the orbiter demonstrated high recession rates, and more than half the mass of ablator heat shield was removed during the 180-second heat pulse. Given the flight experience of Galileo, if the mission were done again, heat shield mass margins would probably grow, leaving less mass for science.

Both the Galileo and Pioneer Venus probes used FM-5055 carbon phenolic for the heat shield material. If possible, it would be prudent to employ the same TPS material for future missions that would encounter severe entry heating environments. Unfortunately, there is a very limited supply of the Avtex precursor rayon needed to make FM-5055 carbon phenolic. Since this material has a strong flight legacy with extensive support through ground-based arc jet tests, it is the only material capable of meeting the failure probability requirements for a Mars sample return to Earth. Unless alternative carbon phenolic materials for the legacy FM-5055 can be qualified for such severe environments, a Gas Giant probe mission will require development and qualification of a new TPS material.

Following completion of the Galileo mission, the Giant Planets Facility at ARC used to qualify the FM-5055 material for flight was decommissioned. As a result, the United States does not currently have any arc jet test capability that can be used to simulate the entry conditions relevant to giant planet entry probes.

3.5 Exploration of Titan

Entry conditions at Titan are relatively benign. Entry velocity is only 6–6.5 km/s, with maximum heating rate of about 100 W/cm2 (up to 50% radiative) and stagnation pressures only about 0.1 atmospheres. Existing arc jet facilities should be adequate for proposed missions to this destination.

The Huygens probe, developed by the European Space Agency, used a tiled heat shield. It was tested in a plasma wind tunnel at the University of Stuttgart, both in a simulated Titan environment (77% N_2, 20% Ar, 3% CH_4) and in a pure nitrogen environment. After the mission was launched, new models for the atmosphere and adjustments to the planned trajectory indicated that heating could exceed the design levels, with a larger radiative contribution. NASA undertook a review of the heat shield, to determine whether the design was adequate and whether modifications to the entry trajectory might improve heat shield margins. Review activities included radiation testing, arc jet testing, and material response modeling. Ultimately, design adequacy was confirmed, but this was an example of the value of arc jet testing, even during mission operations.

3.6 Material Technology Advancement

Flight projects typically select thermal protection materials that have a high Technical Readiness Level (TRL), and cannot afford to invest in material development. Mass-efficient TPS solutions are not available for many of the mission opportunities in the next 30 years and material technology development will be required. Current research projects are employing a functional design approach for materials development. Innovative materials concepts, which involve graded densities, new polymer chemistries, and dispersed additives to improve insulation and radiation properties, are being considered. Since these concepts are new, arc jet testing will be essential for generating insight into material response across a range of mission-relevant environments. As material development proceeds to higher TRLs, advanced thermal protection technologies will reflect a clearer potential benefit as a vehicle subsystem.

Traditionally, low-TRL materials have been subjected to simple screening tests, in which developmental samples are exposed to flight-relevant environments with binary "pass-fail" performance criteria. More than 8000 arc jet tests were performed in the decade leading to Apollo flights, with a large percentage concentrated in the early development phase. It is estimated that modern simulation capabilities can help to reduce that total to around 2000, as shown in Figure 3.4, but flight-relevant testing will remain critical.

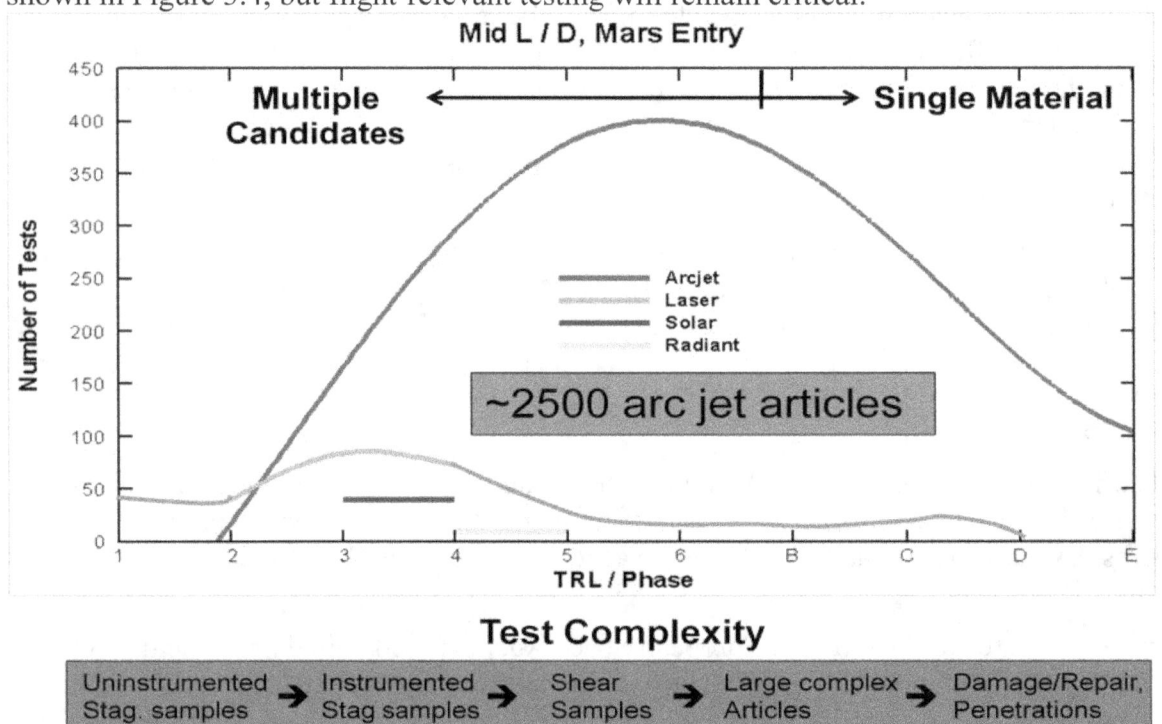

Figure 3.4. Proposed testing requirements for qualification of an ablative TPS
[Wright, Anticipated Arc Jet Usage]

The efficient use of simple, less costly thermal tests combined with more scientific understanding can be used to screen candidates in this manner prior to arc jet testing. New arc jet facilities that provide more precise control and knowledge of the test conditions, and that have enhanced diagnostics to analyze gas species will increase the insight that can be derived from testing early in the development cycle. Instrumentation that records time-accurate surface behavior, recession rates, and species concentrations downstream of the test article are needed to promote understanding of material response and more accurate prediction models. Deeper appreciation of functional performance and failure modes will define the potential of novel materials for mission applications.

In recent decades, new material development has competed for arc jet access with mission-critical subsystem development and sustaining engineering. Without the time-critical delivery requirements of the competing work, material development tests are commonly rescheduled, which complicates planning and milestone performance for these efforts. In a new complex, dedicated facilities should be made available for low-TRL material development activities.

3.6 Summary

Finding 3.1: Arc jet testing will be needed even for LEO return missions, for which materials already exist and operational experience is in hand. Existing test capability is adequate for this mission type. Capacity requirements to support LEO return are dependent on certification philosophy.

Finding 3.2: NASA and DOD share a mutual reliance on arc jet test capability. Air Force strategic planning relies on the availability of NASA arc jets for testing of hypersonic cruise vehicles. AEDC supports high-shear test conditions that are relevant to NASA.

Finding 3.3: Greater arc jet capability will be required for missions that NASA intends to fly within 30 years. Enthalpy, combined convective-radiative heating, test gas, shear, pressure, turbulence, and model size are among facility features to be addressed.

Finding 3.4: Arc jet facilities that provide more precise control and knowledge of the test conditions, and that have enhanced diagnostics to analyze gas species will benefit the maturation of low-TRL thermal protection concepts and materials technologies.

Finding 3.5: The available TPS choices for flight consideration are not optimal for many missions, and the TPS choices could be increased with improved accessibility and lower cost to use arc jets to develop low TRL concepts.

4.0 Current Capabilities and Limitations

This section describes basic arc jet operation (Section 4.1) and then compares the performance capabilities of arc jet complexes that can support larger scale TPS development and certification. Additional description of the arc jet complexes can be found in Appendix F. Not evaluated here are small research and development arc jets and arc jet facilities that are not used for TPS material testing such as the arc facilities at the Langley Research Center, AHSTF, and HyMETS. In addition, foreign (non-US) facilities were not considered as viable alternatives as a primary testing source for many reasons, including ITAR restrictions, control of access, control of schedule, and logistics.

Four arc jet complexes that can support TPS development and certification are:

1. Ames Research Center Arc Jet Complex
    a. Interaction Heating Facility (IHF)
    b. Aerodynamic Heating Facility (AHF)
    c. Panel Test Facility (PTF) [includes the Truncated-PTF (TPTF)]
    d. 2" x 9" Turbulent Flow Duct (TFD)
2. Johnson Space Center Atmospheric Re-entry Materials and Structures Evaluation Facility (ARMSEF)
    a. Test Position 1 (TP-1)
    b. Test Position 2 (TP-2)
3. Arnold Engineering Development Center Arc Jet Complex
    a. High Enthalpy Ablation Test Unit H1 (HEAT-H1)
    b. HEAT-H2
    c. HEAT-H3
4. Boeing Large Core Arc Tunnel Arc Jet Facility (St. Louis)

Comparisons will be made for the various types of testing techniques in the individual arc jet facilities, specifically, free-jet stagnation testing, free-jet shear testing, and panel testing. For each of these three types of testing, comparisons will be shown for test model surface heat flux versus surface pressure. In addition, for stagnation testing, a comparison will be shown for centerline enthalpy versus pressure, and for shear testing, a comparison will be shown for heat flux versus shear force.

Although the vacuum system, high pressure gas system, and cooling water systems are important for determining the complex capability, the power supply size will typically limit the maximum capability of a facility in terms of flow-field size, enthalpy, heat flux, and run time. Therefore, only the power supply capabilities at the various complexes will be compared here.

All of the facility performance data and mission trajectories presented in the figures below were obtained from Howard, Summary of Current NASA Facility Maps; Venkatapathy, Capabilities Needed for Future Venus and Outer Planet Missions; Wright, Anticipated Arc Jet Usage; Raiche, NASA Ames Arc Jet Complex Overview; Kardell, Boeing LCAT Facility Capability; Smith, AEDC Arc Jet Facility Capabilities; Ares, Arc-Heated Test Facility Investment; Shepard and Carlson, Upgrading of NASA-Ames High-Energy Hypersonic Facilities; NASA SP-8014, Entry Thermal Protection.

4.1 Basic Arc jet Operation

Arc jets are one part of NASA's hypersonic thermal testing capability and are used to provide critical data for the research and development of TPS materials and techniques; to qualify, certify, and validate the suitability of TPS materials and processes for flight; and to support TPS damage assessment and the verification of repair techniques. The basic operation of an arc jet, depicted in Figure 4.1, involves using a high-power electric arc to heat a test gas to very high temperature and then expanding and accelerating this heated test gas through a nozzle and onto a stationary test model or calibration probe located in a vacuum environment. This high-energy flow of ionized gases simulates the surface pressure, convective heating, and shear force conditions associated with hypersonic flight, and the vacuum environment allows the high-altitude conditions to be established and maintained.

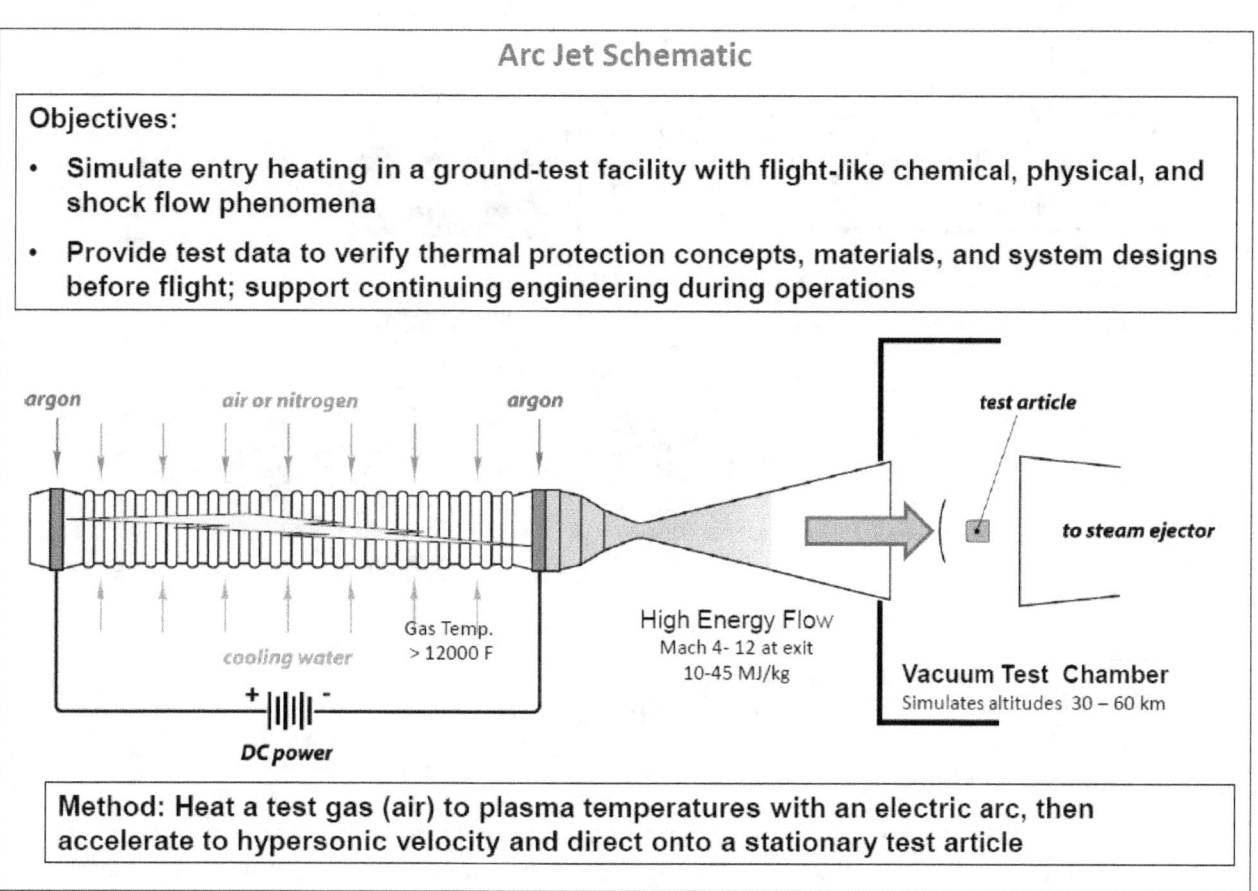

Figure 4.1. Arc jet schematic
[Raiche, ARC Overview]

Several systems are required for an arc jet, as shown in Figure 4.2. The test medium is air, a specialty gas, or a combination of gases, which require pressurized storage and delivery systems. A high-voltage, high-current electrical distribution system is necessary to generate the arc. The arc heater must be actively cooled to protect against the high-temperature conditions, and this is accomplished through a cooling water system. The vacuum in the test chamber is established and maintained through the use of a steam ejector system or mechanical pumps. The steam ejector system uses a boiler to generate the steam. Some facilities have pollution control systems which require capture and condensing of the exhaust products from the ejector system. The test article is inserted into and retracted from the test stream through a model support system. All of these tunnel systems are operated through a facility control system. Test chamber conditions and model reactions are captured and processed through a data acquisition system that collects and stores the raw data and then reduces it to usable information.

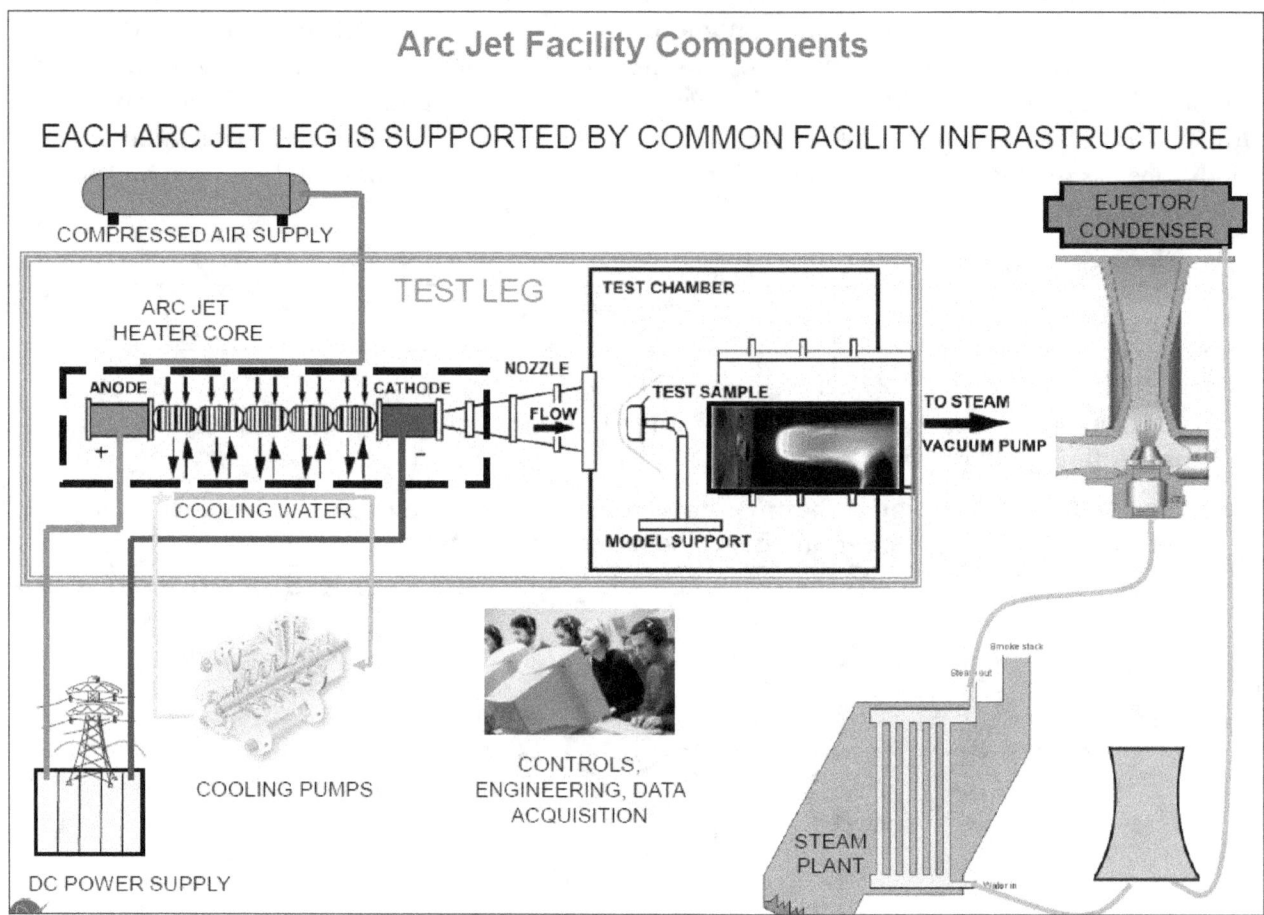

Figure 4.2. Arc jet facility components
[Raiche, ARC Overview]

4.2 Simulation Regime

The overall arc jet complex performance is shown in Figure 4.3 as a function of simulated altitude and velocity for stagnation-type testing. Several system entry trajectories are also shown on the plot for the Shuttle (similar to Crew Exploration Vehicle (CEV) and International Space Station (ISS) return), the Apollo (similar to CEV Lunar return), a Mars return, a far solar system probe return, and an Intercontinental Ballistic Missile (ICBM). In addition, the peak heating point for each mission trajectory is shown. Notice that the present facility envelopes do an adequate job of covering a LEO entry (Shuttle return) and almost capture the Apollo Lunar return peak heating point; however, they are inadequate for simulating the peak heating regions for Mars and far solar system returns. The present facilities are also inadequate for simulating other planet entry peak heating points, such as Saturn and Jupiter.

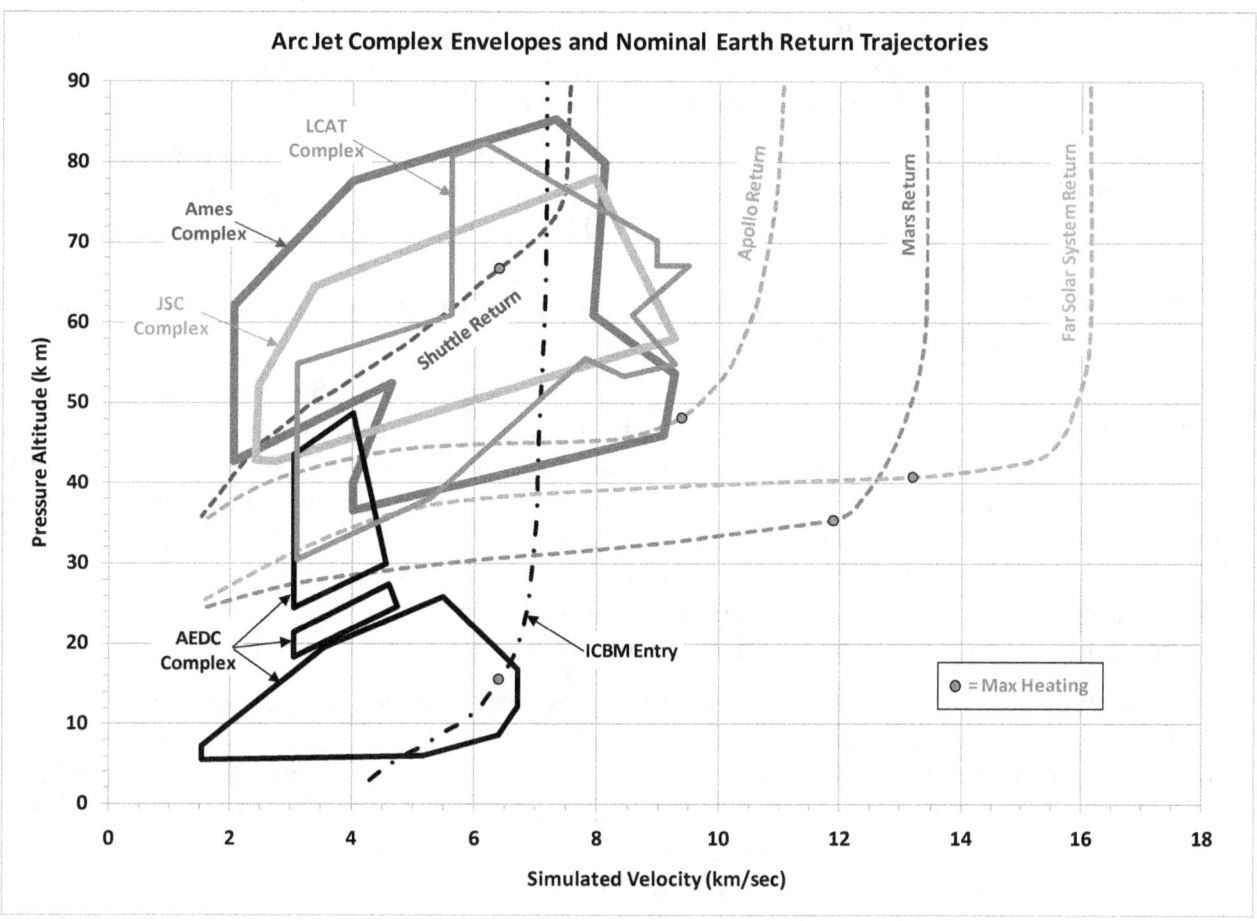

Figure 4.3. Comparison of NASA, AEDC, and Boeing arc jet facility performance with several nominal Earth return trajectories.

The ARC, JSC, and LCAT simulation envelopes are at the higher altitudes, which capture the major portion of the Shuttle and CEV trajectories, while the AEDC facilities are primarily in the lower altitude regime, suitable for DOD weapon trajectory simulation, and capture the maximum heating point of an ICBM entry. However, the AEDC H2 facility does have a region of interest for NASA missions: it captures a portion of the CEV (Apollo), Mars return, and far solar system return trajectories that are generally associated with higher pressure and shear conditions.

Notice that the ARC facilities' combined simulation envelope is larger and encompasses most of the JSC simulation envelope. Also notice that the LCAT facility has a significant overlap with the NASA facilities and the AEDC H2 facility. The AEDC H2 facility captures only a small portion of the simulation envelope that is relevant to NASA missions; therefore, most of the discussion in the following sections will focus primarily on comparing the capabilities of the three complexes that are of most interest: ARC, JSC, and Boeing LCAT. In addition, the H1 and H3 facilities at AEDC do not have any regime that is of interest to NASA missions, and these two facilities will not be considered further.

In the sections below, the performance capabilities of each complex will be compared for three different types of testing: stagnation, shear, and panel.

Stagnation-type testing generally uses a circular model with a flat face, rounded face (ISO-Q), or hemispherical shape. In addition, stagnation-type testing can be performed on special features such as pressure ports in TPS, nose tips, or leading edges. The model is usually positioned in the free-jet downstream of the nozzle so that it experiences stagnation flow on the region of interest. Facilities that generally perform stagnation testing are TP-2, IHF, AHF, LCAT, and H2.

Shear testing generally uses wedge type models that are injected into the free-jet flow downstream of the nozzle exit. The test samples are usually flat material panels that can be fixed to the wedge surface. The wedge angle of the model holder is tailored to provide a specific heating rate, pressure, and shear force on the test sample surface. Facilities that generally perform shear testing are the same as the facilities that perform stagnation-type testing: TP-2, IHF, AHF, LCAT, and H2.

Panel testing generally involves larger models than either stagnation or shear testing and generally involves testing of special features in the TPS, such as gaps or seams between TPS blocks or tiles or protuberances on the external mold line. There are two basic types of panel test facilities. One type uses a channel flow where the test article makes up one wall of the channel. It is usually difficult or impossible to have visual access to the sample during the test in a channel-type facility. In addition, the test sample has to be relatively flat to conform to the channel profile; therefore, these facilities are not generally used to test outer mold line protuberances. Facilities that perform panel testing using a channel flow are TP-1 and the TFD.

The second type of panel testing facility generally uses a semi-elliptic nozzle to expand the flow as wide as possible to cover a larger test surface, and so the nozzle has a flat lower surface. The test article is attached to the lower, flat surface at the exit of the semi-elliptic nozzle and can be pitched into the flow. This type of panel facility has good optical access during the test, can test protuberance-type of features, and can change pitch, even during the run to provide a larger range of test conditions. Facilities that perform this type of testing are the PTF and IHF.

4.3 Stagnation Testing

Stagnation performance testing envelopes are shown in Figure 4.4 for a 10-cm (~4-in.) hemisphere for the IHF, AHF, TP-2, and LCAT in terms of cold wall heat flux on the model surface and model surface wall pressure (stagnation pressure). Note that a 10-cm hemisphere size is used only as a comparison between facilities. One of the LCAT nozzles can only support a 8.9-cm (3.5-in.) diameter model; and therefore, a portion of the LCAT envelope, shown in Figure 4.4, cannot support a model as large as 10 cm. Figure 4.5 shows stagnation performance in terms of centerline enthalpy and model wall pressure. Similar to the overall comparison envelopes shown in Figure 4.3, the ARC facilities, IHF and AHF, encompass most of the JSC TP-2 performance envelope and a significant portion of the LCAT envelope. Two trajectories are shown in the two figures below for reference: a Shuttle return (ISS return) and an Apollo return (Lunar return). In addition, the peak heating conditions are also shown in Figure 4.4 for a Saturn aerocapture and Saturn hyperbolic Entry. Present arc jet facilities are inadequate for simulating Saturn and other Gas Giant entries.

In general, the NASA facilities and the AEDC facility can test larger models than the Boeing facility because they have a family of larger nozzles. However, flow-field size for a given condition is determined by arc heater power. Therefore, if the power delivered to the test gas is higher, the flow field can be expanded to a larger size for a specific test condition, assuming larger nozzles are available.

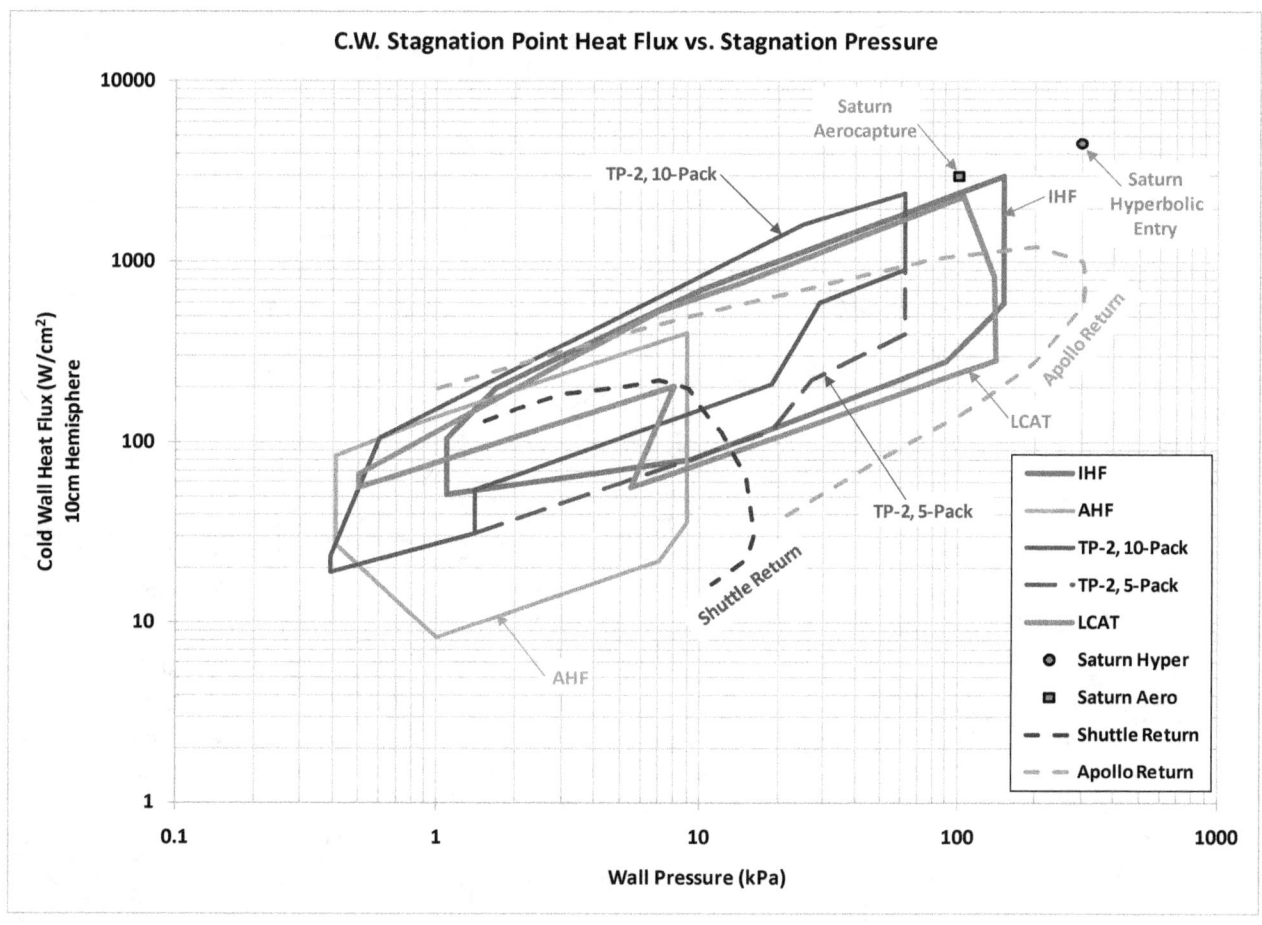

Figure 4.4. Stagnation point heat flux vs. stagnation pressure for a 10-cm hemisphere

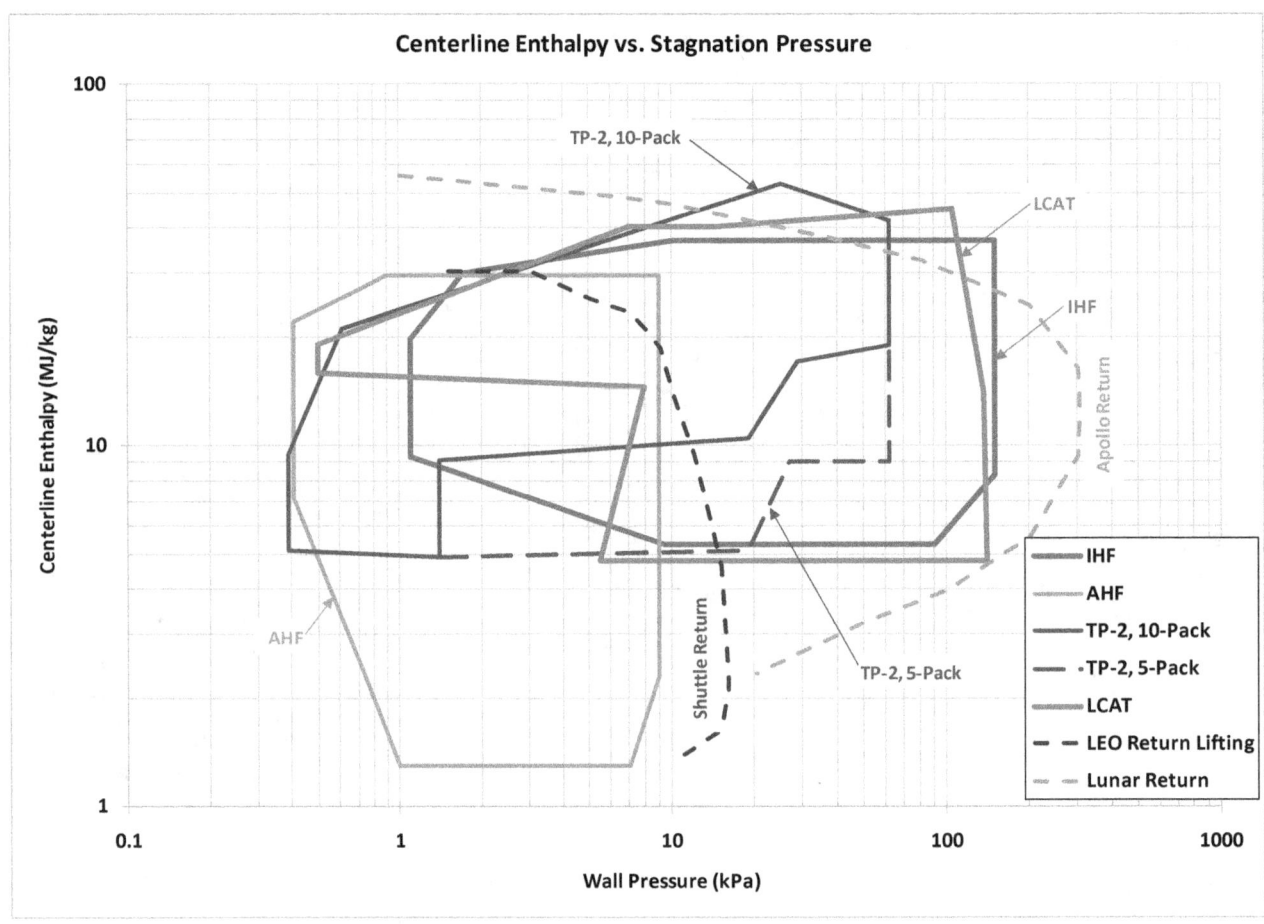

Figure 4.5. Stagnation facility comparison, centerline enthalpy vs. wall pressure for 10-cm hemisphere.

4.4 Shear Testing

Free-jet shear testing is primarily performed using a wedge model. The wedge angle can be varied to acquire the desired heat flux, pressure, and shear on the test article. Figure 4.6 and Figure 4.7 show comparisons of shear testing among various facilities. Notice that the NASA and Boeing facilities have a considerable overlap, while the AEDC facility can obtain higher pressures and shear forces typically required for weapons systems testing. The TP-2 facility is not shown in these figures, but is expected to have an envelope slightly smaller than the IHF facility.

Free-jet shear testing differs from panel testing in the area of test conditions and sample sizes. The test conditions, pressure, heat flux, and shear are higher for free-jet shear testing than for panel-type facilities; however, the sample sizes are typically smaller. Notice that the sample sizes shown in figures 4.6 and 4.7 are on the order of 4 inches by 4 inches where panel samples are much larger. The TPTF, although technically a panel test facility, is included with the shear test facility envelopes in the figures below, because it has sample sizes and test conditions comparable to the free-jet facilities.

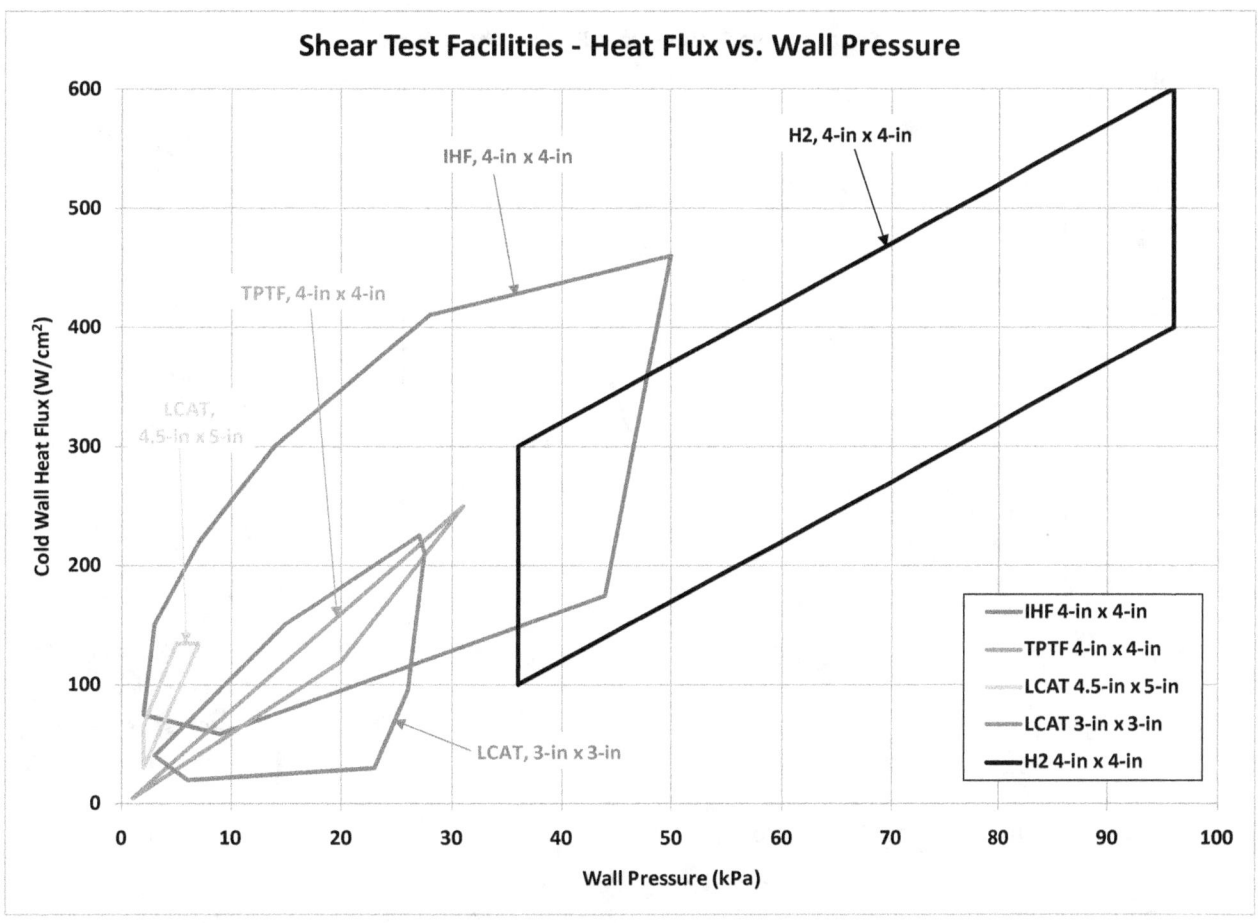

Figure 4.6. Shear test facility comparison, cold wall heat flux vs. wall pressure

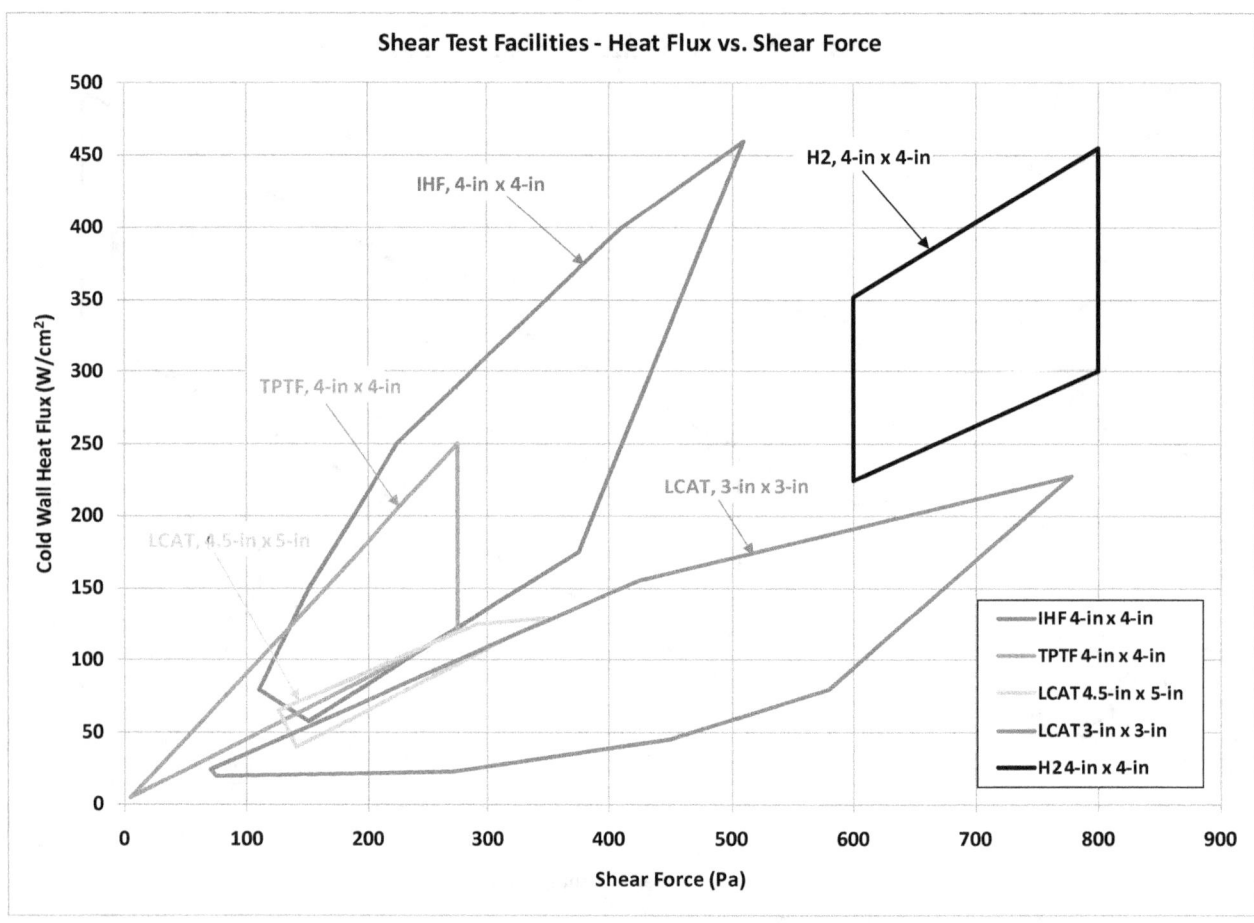

Figure 4.7. Shear test facility comparison, cold wall heat flux vs. shear force.

Figure 4.8 shows the shear test performance for the facilities shown in the figures above compared with three trajectories: ISS direct return, Lunar direct return, and Mars direct return. All three of these are ballistic trajectories with full margins applied. Notice that the facilities adequately cover an ISS return, but are not adequate for simulating a Lunar or Mars return.

The MSL test program used arc jet facilities at JSC, ARC, AEDC, and Boeing to perform shear testing. None of the facilities could adequately match all of the required testing regimes of interest; however, among all of the facilities, a sufficient coverage of the conditions was accomplished to evaluate the TPS under appropriate shearing conditions.

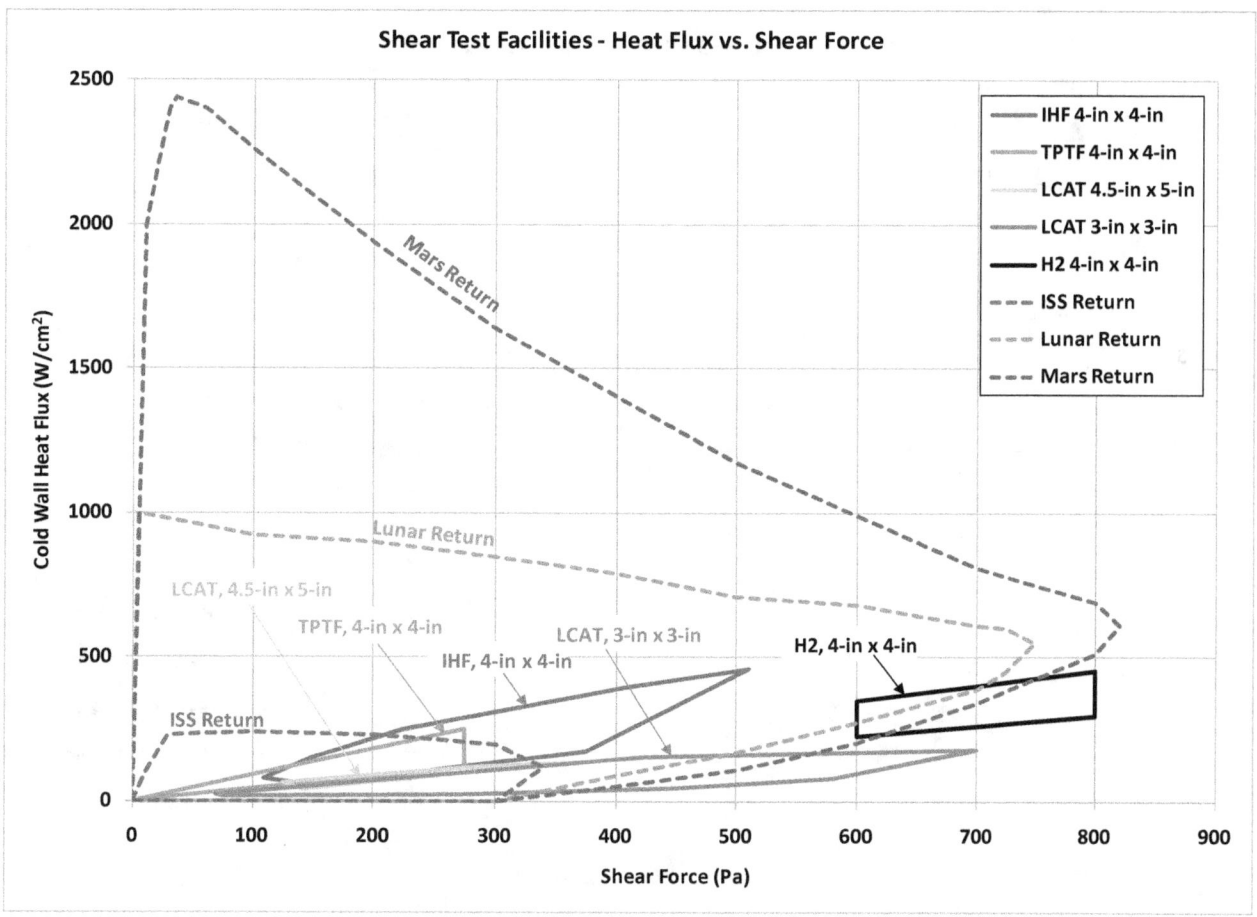

Figure 4.8. Shear test facility capability compared with several ballistic trajectories (trajectories shown with full margins).

4.5 Panel Testing

Four facilities in NASA are specifically configured for panel testing. They are the JSC TP-1 and ARC TFD, which are both channel flow facilities, and the IHF and PTF, which are configured with semi-elliptic nozzles and test plates, which can be pitched into the flow. Figure 4.9 and Figure 4.10 show the performance capability of the four facilities in terms of heat flux versus pressure. Figure 4.10 is the same as Figure 4.9 with the addition of the TFD. Not shown is the new 4-inch by 4-inch test capability in the TP-1 facility, which should have higher heat flux and pressure similar to the TPTF shown in figures 4.6 and 4.7.

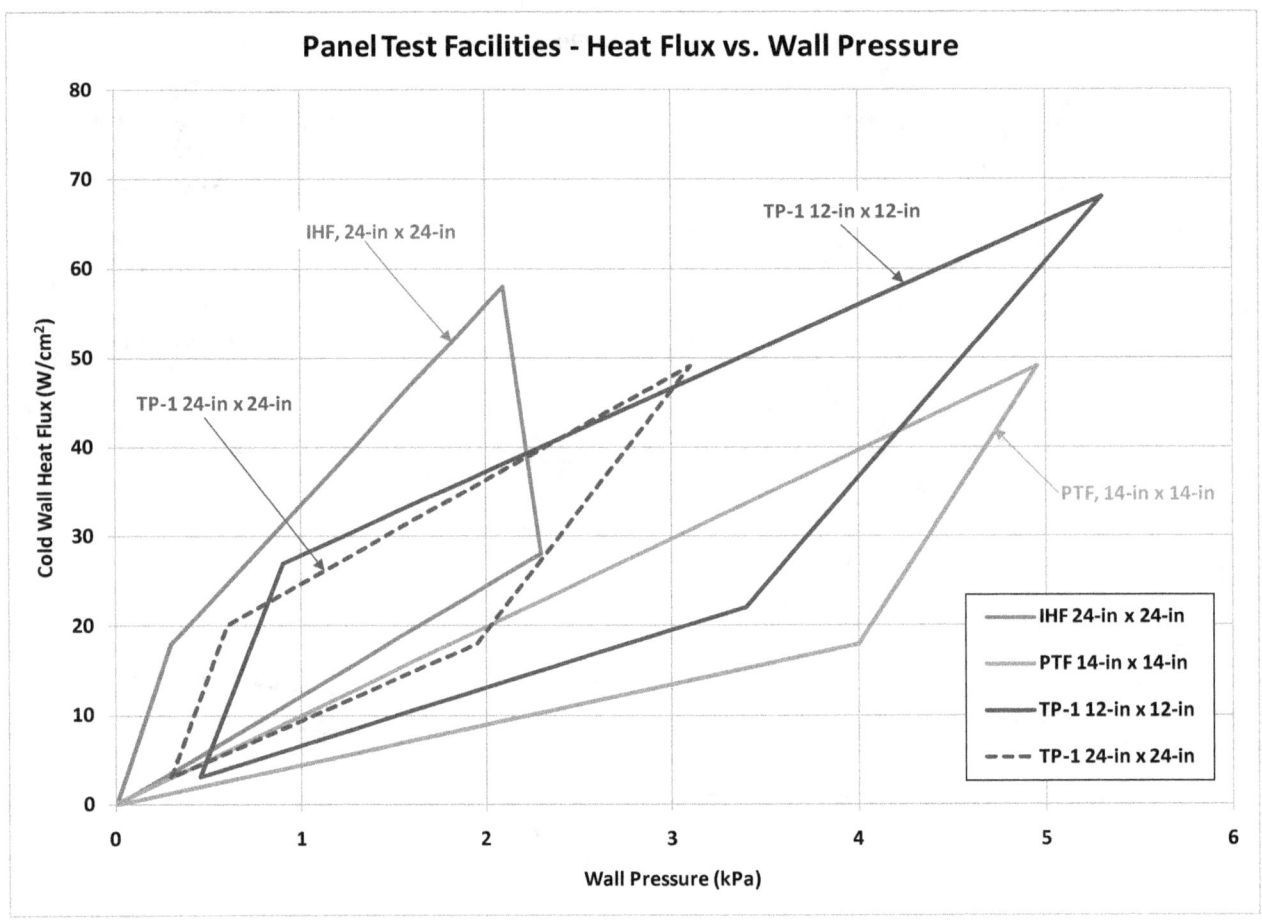

Figure 4.9. Panel test capability showing heat flux vs. pressure.

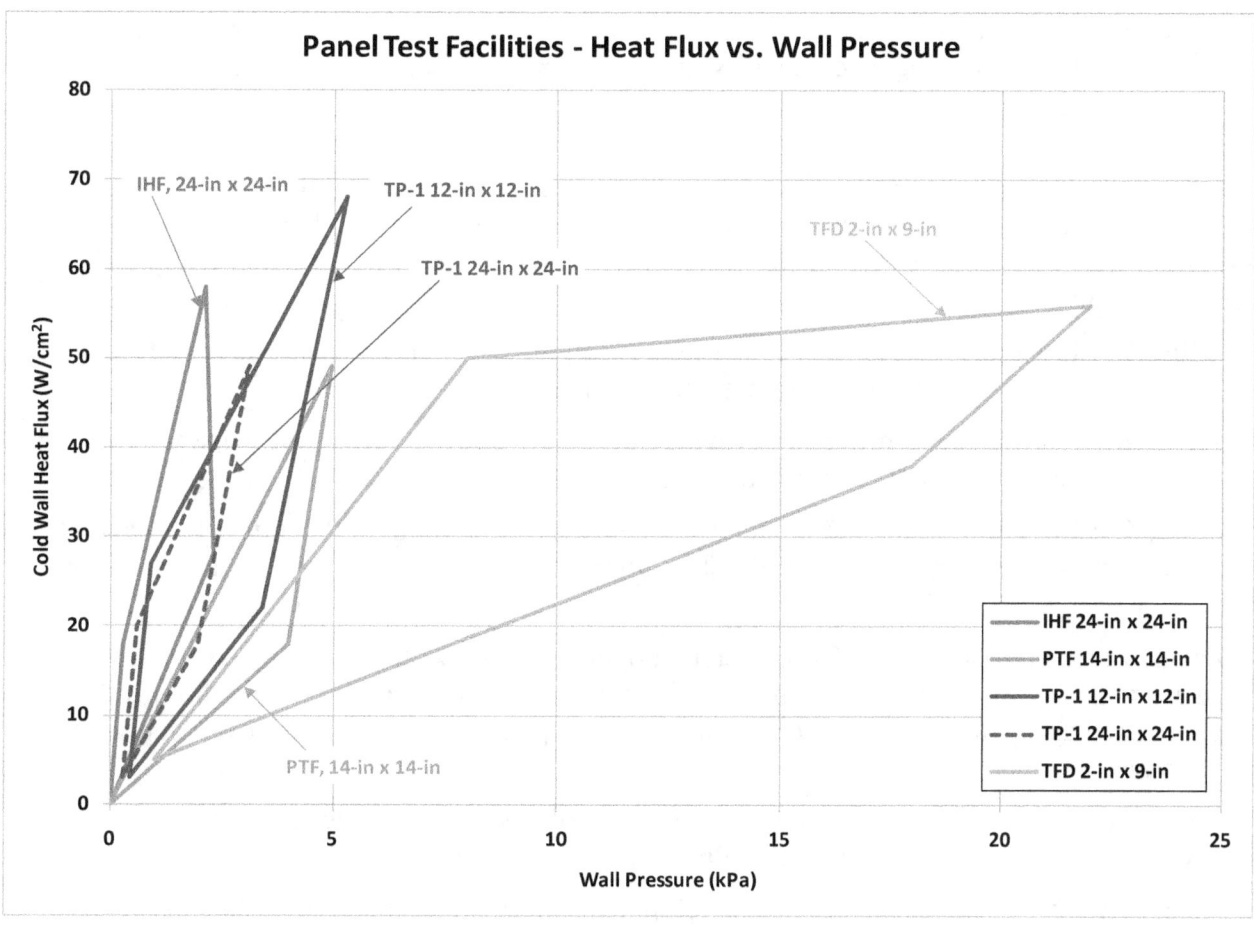

Figure 4.10. Panel test capability showing heat flux vs. pressure with addition of the TFD.

4.6 Power Supply

The larger of the two power supplies at ARC, which is rated at 60 MW continuous operation, can also operate at 75 MW for 30 minutes and 150 MW for 15 seconds. However, the power supply is presently operating under reduced capacity because of a fire which destroyed one of the six rectifier modules. This reduction in capability does not affect performance of the IHF facility, as it runs at power levels below the maximum capability. The JSC power supply is rated at 10 MW continuous operation and can operate at 12 MW for 15 minutes. The AEDC power supply is rated at 60 MW continuous operation and has operated at 77 MW for approximately one minute. The LCAT facility has a 12 MW power supply.

The ARC and AEDC power supplies are both 60 MW continuous output-rated power supplies; however, the ARC power supply is configured to operate at higher current and lower voltage (which is suitable for low pressure flow environments), while the AEDC power supply operates at higher voltage and lower current (which are suitable for high pressure operations). The different configurations of the ARC and the AEDC power supplies is significant if consideration is given in the future for installing a new arc heater capability at AEDC to operate in the range of interest for NASA missions. The AEDC power supply cannot be reconfigured to operate at low pressure conditions (high current and lower voltage) and, therefore, a new power supply would need to be procured if a NASA-type heater were to be operated at AEDC. The JSC and LCAT facilities are both much lower in power output. This will be a consideration if upgrading an existing facility or installing a new facility in an existing complex is considered.

4.7 Summary

Finding 4.1: The simulation regimes of the present facilities are sufficient for simulating low Earth orbit entries and certain regimes for Lunar return. However, these facilities are inadequate to fully evaluate the TPS for a Mars or deep space Earth return or certain planet entries such as the Gas Giants.

Finding 4.2: Existing NASA facilities have capability in terms of model size and performance that is not duplicated by DOD or commercial facilities.

Finding 4.3: An inspection of the performance envelopes shows that the JSC capability is basically a subset of the ARC combined capability.

Finding 4.4: Incremental improvements in arc heaters will not deliver the higher test capability needed for more ambitious NASA exploration missions to Gas Giants and human return from Mars.

Finding 4.5: The AEDC power supply cannot operate at low voltage and high current, which is required to power a NASA-type arc heater which operates at low pressure. Therefore, if AEDC is considered as a potential site for operating NASA-type arc heaters, a new power supply will need to be procured.

Conclusion 4.1: NASA must bear the responsibility of designing and developing facilities with improved capability in order to successfully support future NASA missions.

Conclusion 4.2: A focused technology development effort is required to establish the arc jet testing capability required by NASA for future missions.

5.0 Arc Jet Capacity

Section 5.1, Workforce, compares the workforces at ARC, JSC, and AEDC arc jet complexes. Section 5.2, Throughput, evaluates the operations at the ARC, JSC, AEDC, and LCAT facilities, the throughput, the customer base, and types of testing performed by these customers. The proposed LEAF facility is considered in the context of capability gaps for mission requirements in Section 5.2. Section 5.3 considers the condition of the arc jet infrastructure. Section 5.4 evaluates mothballing existing facilities. Section 5.5 identifies options for increasing throughput.

5.1 Workforce

Arc jet facilities require a workforce with unique skills and qualifications. Both the ARC and JSC arc jet test facilities use segmented arc heaters, which are considered to be state of the art. The operation and maintenance of segmented arc heaters at both facilities is complex and requires a staff with special knowledge and skills. Detailed attention must be given to the assembly of the segments, insulators, tie rods, water, and air hoses to insure safe and reliable operation. The high pressure air and cooling water supply systems are relatively standard at both facilities, but the water must be de-ionized to reduce electrical conductivity and scaling within the heater.

Arc jet power supplies operate at very high AC voltages and currents that must be rectified to provide necessary DC current and that require special current controls to maintain stable arc operation. A reliable operation of high power electrical systems requires special attention to maintenance, repair, test, and safety procedures.

Both the ARC and JSC arc jet facilities operate with dedicated boiler and steam systems to generate the vacuum pressures needed for simulating entry conditions. The steam vacuum systems for arc jet facilities must be well maintained to ensure safe and efficient operation. These facilities also produce gas flows with extremely high temperatures and heat transfer rates, which require special knowledge of handling hot gas flow to prevent facility damage. Unique high-temperature instrumentation is necessary for test model measurements, and this includes non-intrusive instrumentation often needed to characterize the flow field.

Arc jet operation personnel must be knowledgeable about the special procedures and precautions required to ensure both personnel and plant safety. Safe operation of these systems and subsystems requires a staff with special qualifications and training. Extensive training and experience is necessary to be qualified for performing required tasks. The special skills and knowledge required have been gained through years of operational experience, teaming, and continuity of training. Continuity for training new facility personnel is necessary for successful operation.

Simply relocating existing personnel is not always possible, and the availability of workforce should be considered when choosing a site for a new construction facility.

While a workforce with special skills is necessary, the staff need not be large. Both ARC and JSC have the appearance of excess staffing at both facilities. Since operational infrastructure support systems (power, steam, coolant water, and gas supply) are common to all test units, an operating crew of 15 to 20 people should be sufficient staffing for a single shift operation. The primary responsibility of this test operations crew is to support the day-to-day test programs in the arc jet facility.

Other activities associated with an arc jet facility, but which are not part of the daily operations, would include arc jet technology and development, hardware design, fabrication services, plant maintenance and repairs, minor facility modifications, procurement, safety, configuration management, and quality assurance. The personnel who perform these tasks may or may not be dedicated to the arc jet facility. Preferably, some of the more closely related activities such as arc technology and facility design would be staffed by full-time dedicated personnel. Other arc jet facility support of a less technical nature may be accomplished by personnel who also support other Center test facilities. The cost of these services may be shared, thus reducing the overall budget required for these services.

The ARC arc jet staff is composed of 20 test operations personnel for day-to-day activities, which include model preparation, arc heater and test cell readiness, pre-operational set up, test, and post-operational efforts. Of these 20 people, 5 are civil service employees and 15 are contractor employees. This level of staffing is reasonable for a single-shift operation. There are 33 personnel providing operational support. This support includes engineering (design, drafting, documentation, project support, and quality assurance), maintenance and repair (plant maintenance and fabrication services), and administrative services (safety, configuration management, procurement, and contract support). Resources could be saved by allowing administrative and maintenance personnel to share their time with other Center facilities. This would allow these personnel to charge to arc jet facility projects only when they are provide direct support. At other times, these personnel would be available for other Center tasks and should charge to these tasks. An independent review of the entire arc jet operation can reduce staffing levels by up to 30%.

The JSC staff has 3 civil service employees and 33 contractor employees in a two-shift operation. The contractor staff is evenly divided between engineers and technicians. The JSC arc jet facility is presently booked full time with Space Shuttle testing. Once the Shuttle tests have ended, the JSC arc jet facility should return to a single-shift operation. This would allow the contractor staff to be reduced by about one-third [Madden, JSC Arc Jet].

The operations workforce at AEDC consists of 20 personnel plus 2 engineers supporting the arc facility technology development projects. In addition, high-pressure air and the mechanical vacuum source are provided through AEDC's central base support.

5.2 Throughput

5.2.1 Ames Research Center

The ARC arc jet complex comprises four tunnels: the 20 MW AHF, the 60 MW IHF, the 20 MW PTF, and the 12 MW TFD. The ARC arc jet complex has a staff of 53; and for workforce planning, uses a single-shift operation 40 weeks per year, or 160 available calendar days. Multiple facilities can operate on the same calendar day, which gives the facility manager flexibility in scheduling and accomplishing revenue-generating Occupancy Days (OD). ARC targets 220 ODs per year for operations and 2 facility runs per OD. Thus, a single calendar day can generate more than one OD, and approximately 50% of the operational days for the AHF and IHF yield two or more runs per day.

Test operations are conducted Monday through Thursday for nine hours each day. The first Friday of the two-week schedule is reserved for maintenance, and the second Friday is utilized for alternative work schedules. For high-occupancy periods, extended shifts and weekend maintenance are used.

For the last five fiscal years, ARC's throughput was as shown in Table 5.1.

Table 5.1. Throughput for ARC Arc Jet Complex
[Raiche, ARC Overview]

	Test	Calibration	Charged	Actual	AHF		IHF		PTF		TFD	
Year	Models	Probes	OD	OD	Models	Probes	Models	Probes	Models	Probes	Models	Probes
FY05	555	460	175	215	220	289	335	170	0	1	0	0
FY06	292	434	135	205	43	144	219	261	26	23	4	6
FY07	404	748	218	265	209	445	125	223	68	25	2	55
FY08	423	786	171	242	223	437	194	332	0	9	6	8
FY09	404	435	143	206	139	214	171	197	94	24	0	0
Totals	2078	2863	842	1133	834	1529	1044	1183	188	82	12	69
Averages	416	573	168	227	167	306	209	237	38	16	2	14

Data for test models, calibration probes, and Occupancy Days (OD) is tabulated per fiscal year (FY)

The table captures the variability in annual hypersonic thermal testing requirements. The AHF and IHF are the most utilized of the four facilities, and these two facilities are approximately equally active and account for over 90% of the test models and calibration probes through the five-year period. Approximately 40% of the testing is for models and 60% is for calibration probes. The number of calibration probes was unusually high during this period because of the Orion Project's need to establish new operating conditions in the facility. To provide a look at recent users, FY08 and FY09 utilization was approximately 80% NASA (4% Aeronautics Research Mission Directorate (ARMD); 6% Space Operations Mission Directorate (SOMD); 20% Science Mission Directorate (SMD); and 50% Exploration Sciences Mission Directorate (ESMD)); 12% other U.S. Government; and 6% from the commercial sector. This distribution is graphically represented in Figure 5.1.

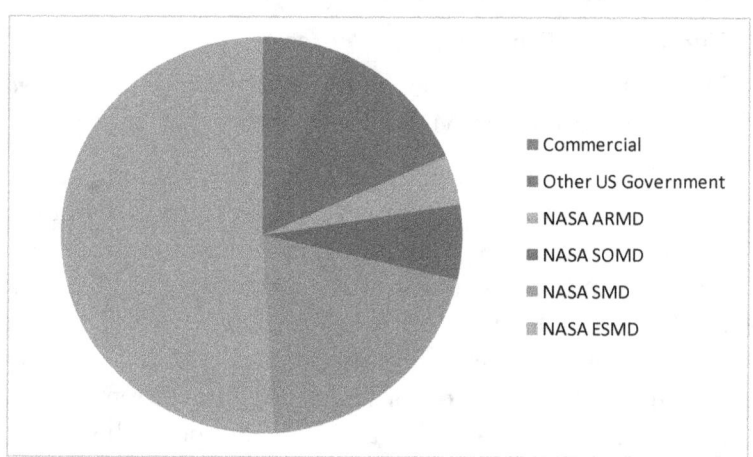

Figure 5.1. ARC arc jet complex customer base

Not captured in Figure 5.1 is the significant increase in the commercial sector percentage from FY08 to FY09.

5.2.2 Johnson Space Center

The ARMSEF comprises the channel nozzle TP-1 and the conical nozzle TP-2; both are 10 MW facilities. The ARMSEF has a staff of 36 personnel working two shifts, 5 days a work week. The staff operates both the ARMSEF and the separate Radiant Heat Transfer Facility (RHTF). The simultaneous operation of TP-1 and TP-2 is not possible because of electrical power and workforce constraints. It is possible to operate either TP-1 or TP-2 and RHTF at the same time, though, but this is not done due to workforce constraints.

For the last four calendar years, JSC's throughput was as shown in Table 5.2:

Table 5.2. Throughput for JSC Arc Jet Complex
[Del Papa, ARMSEF and Madden, JSC]

Year	Tests	Test Models	Calibration Probes	TP-1 Models	TP-1 Probes	TP-2 Models	TP-2 Probes
2006	229	143	86	35	11	108	75
2007	158	94	64	29	5	65	59
2008	192	141	51	19	5	123	46
2009	187	128	59	8	8	120	51
Totals	766	506	260	91	29	416	231
Averages	192	127	65	23	7	104	58

Data for test models and calibration probes is tabulated per calendar year

The TP-2 is used more than TP-1 and accounts for 85% of the models and probes tested through the four-year period. This is due to the prime customer being the Shuttle Program, and panel testing of tiles has been more critical than stagnation testing to Shuttle-related activities. Approximately 65% of the testing is for models and 35% is for calibration probes. The ARMSEF is overwhelmingly dedicated to the Shuttle and Constellation Programs.

5.2.3 Air Force Arnold Engineering Development Center

The arc jet complex at AEDC comprises three tunnels: the 30- MW H1 segmented arc heater, the 42 MW H2 arc-heated wind tunnel, and the 70 MW H3 arc heater.

The AEDC complex has a staff of 20 plus additional personnel for support systems. Test operations are conducted through a single shift, 40 hours per week, and 50 weeks per year. AEDC utilizes overtime as needed to maintain schedule, and annual overtime requirements do not exceed 10-15% of the standard shift work hours. AEDC accomplishes 60-90 runs in a typical year, which equates to 300-400 test articles per year (each run involves 4-7 test articles). Eighty percent of the runs are test articles and 20% are calibration probes. Approximately 80% of the runs are for reimbursable customer runs, 15% are for technology development, and 5% are for facility checkout and validation. The arc jet complex at AEDC supports the Air Force, Army, Navy, the Defense Advanced Research Projects Agency (DARPA), the Missile Defense Agency, and NASA (SMD and ESMD).

5.2.4 Boeing Large Core Arc Tunnel

The 5 MW Boeing LCAT is a limited-operations facility with a test staff of four, plus additional personnel who provide maintenance and operational support. The arc jet facility is available during normal working hours. The number of tests Boeing accomplishes each year is a highly variable number ranging from 50 to 1000, and approximately 65% of the testing is for models and 35% is for calibration probes. The facility occupancy is cyclical, but typically ranges from about 25% to 100% with an average availability of about 50%. When not engaged with arc jet testing, the facility staff supports other projects. Boeing projects account for 20-30% of the facility occupancy. The facility also supports NASA, the Air Force Research Lab, and small business innovation research contractors.

5.2.5 Testing Costs

Comparing costs and throughput provides insight into the cost of testing at the ARC, JSC, AEDC, and LCAT facilities. Table 5.3 [deleted from public version of report] contains information on annual operations and maintenance costs and average throughput used to calculate an average cost per test at these four complexes. This cost is not what a customer will pay; rather, it is the average cost the combined funders (the owning organization, any subsidizers, and all customers) pay per test. To provide a more consistent comparison with ARC and JSC, the "Annual Cost" and "Associated Personnel" numbers for AEDC and LCAT are adjusted to capture the additional personnel that provide engineering, maintenance, and operational support. It is not surprising that the average cost per test at the Boeing LCAT is lower than at the Government facilities. Worth noting is the value used for the number of tests (200) at LCAT is conservative, based on Boeing's description of the tests varying from 50 (described as "very lean") to 1000 ("busy"). If a higher number for tests were used, it would decrease the cost per test at LCAT.

5.2.6 Lunar Environment Arc Jet Facility

The Orion Project team undertook the TPS Advanced Development Project Analysis of Alternatives (AoA) Study to determine the most cost-effective means to meet the CEV heat shield certification needs. Their solution was the Lunar Environment Arc Jet Facility (LEAF) Project, a new arc heater with radiant lamp augmentation added to the NASA Ames arc jet complex.

To certify the heat shield for lunar direct return entry velocity, a ground test program must demonstrate the TPS material and system performance and validate thermal response models across a range of flight-like aerothermal conditions. General CEV Lunar return certification needs include 300 W/cm^2 radiant heating capability combined with 400 W/cm^2 of convective heating, shear test capability at heat flux levels greater than 350 W/cm^2, turbulent heating capability, and relevant system responses from panel and integrated system test articles for features on the order of 30 cm.

The LEAF Project was established to develop a combined electric arc jet-heated blow down wind tunnel and radiant heating system capable of simulating much of the aerothermal environment that Orion would experience during returns into the Earth's atmosphere. The general requirements were high radiant and convective heating rates over large stagnation and panel test specimens and combined high heating, high shear, high enthalpy, and low pressure environments. The design included a 75 MW segmented arc heater with conical and semi-elliptical nozzles for stagnation and wedge testing, an independent radiant light source for coupled convective-radiant heating, and installation in a vacant test bay connected to the existing power, vacuum, air, and water infrastructure to minimize initial costs.

LEAF is a technical capability enhancement funded by the Constellation Program with technical requirements determined by and specific to the Orion Project. In the opinion of the AJEWG, the LEAF Project raises two important issues for the Agency. The project analysis indicated that an arc jet capability greater than that which currently exists is needed to meet the requirements of the CEV. And the LEAF Project also illustrates a piecemeal approach to meeting program requirements that does not serve the Agency. Programs are empowered to construct capability without regard to total facility life cycle support, cost, and the impact of that capability on the institution. The narrow focus on Orion requirements for the LEAF facility also allowed no cost latitude for enhancements that could address requirements from other missions. The Preliminary Design Review (PDR) was held in September 2009, the Critical Design Review (CDR) is scheduled for June 2012, and LEAF is scheduled to be operational in October 2015.

5.3 Infrastructure Condition

The condition of the equipment and infrastructure that comprise and support the ARC arc jet complex and the JSC ARMSEF is typical of research systems that were installed in the 1960s and 1970s and have experienced uneven maintenance and recapitalization funding. It is a mix of new, old but maintained, and end-of-life equipment with data acquisition and facility automated control systems ranging between state of the art and (not so recently) obsolete. As will be seen, several systems and components in the ARC and JSC complexes are at or beyond their nominal service life. But this is typical for Government technical facilities; these systems and components are not considered to be in critical condition. The revitalization plans proposed are justified by reliability improvements and meeting expected regulatory constraints and are not by risks to basic operation. It should be noted that there is no commercial supplier base for integrated arc jet service solutions, and that arc jet requirements are unique, they involve high energy, and the tolerance for mistakes is low.

5.3.1 Ames Research Center

The deferred maintenance for the arc jet complex is ~$67 million, and the overall health and estimated life expectancy of the ARC arc jet complex is shown in Table 5.4.

Table 5.4. ARC Arc Jet Complex Subsystem Summary
[Raiche, ARC Overview]

Arc Jet Subsystem Summary and Status

Component	Function	Installation	Major Upgrade	Overall Health	Life Expectancy	Need
20 MW HV Power Supply	DC power for arc heaters	1962	2009	Excellent	>20	Routine maintenance
HP Air Compressors	Test Air Compression/ Storage	2008	-	Excellent	>20	Routine maintenance
Model Insertion Systems	Support model in flow	1961	2000/2010	Excellent	>20	Routine maintenance
Arc Heater Components	Energize and contain plasma	Various	Ongoing--wear items	Good	<10	Spares replenishment
150 MW HV Power Supply	DC power for arc heaters	1974	2008	Good	<10	Transformer rewind/module replacement
Control Systems	Control power/test gas/interfaces	1961/1995	2000-ongoing	Good	<5	Modernization
DI Cooling System	Arc heater/SVS coolant	2007	-	Good	>10	Routine maintenance
SVS Ejectors/ Condensers/CTs	Vacuum source for test chambers	1961	2009/2010/2011	Fair	<10	Structural/seismic reinforcement, modernize controls
Boiler-Steam Generator	Steam for SVS	1961	1990/2011	Fair	<10	Replace
Test Chambers	Vacuum-heating test sections	Various	Various	Fair	<10	Structural reinforcement, wear service
Heat Exchangers	Shield SVS from excessive heat	1961	Various	Fair	<10	Replace with improved design
Plenums/Ducts	Vacuum and exhaust	1961/1975	-	Fair	<10	Structural/seismic reinforcement
Pollution Control System	SVS/Heater Exhaust NOX control	1975	2011	Poor	<5	Replace

Several systems and components in the table are near or at their nominal service life. This is typical for Government technical facilities; these systems and components are not considered to be in critical condition. This analysis was endorsed by ARC in the March 18, 2010, presentation to the AJEWG.

ARC tracks unplanned downtime, and this measure can provide a basis from which facility condition can be inferred. In FY08, ARC experienced 684 hours of unplanned downtime in the AHF and IHF, which translated into an availability of 85% and 82%, respectively. For the first quarter of FY09, the numbers for the AHF and IHF were 36 hours of combined unplanned downtime and an availability (how often the facility is operational when planned) of 90%.

ARC has developed an investment strategy for the arc jet complex, and the objective for the strategy is to achieve an availability of 95%+. This plan identifies: (a) investments to reduce deferred maintenance and maintain operational capability; (b) future infrastructure refurbishment which emphasizes maintainability and robustness; (c) designs to improve energy and resource efficiency; and (d) compliance with all current air and water environmental regulations. This plan integrates a recent facility condition assessment, which identified $5.4 million in reliability and safety issues that need to be addressed quickly and another $15 million that should be addressed within 4 years. These reliability and safety recommendations focus on the preservation of existing operational capability, and not on design changes to significantly improve reliability or to provide an upgrade to technical capability. In addition, capital investment projects identified to address issues with large systems and increase systems capability are as follows:

- Replace Steam Vacuum System (SVS) Boiler. The current boiler was fabricated in 1946 and installed in 1961, is obsolete, has low efficiency, and will not meet expected future NO_x emissions regulations. The ARC boiler is currently operating in compliance with Bay Area air quality limits of 30 ppm for NO_x emissions, but these emission limits are expected to be lowered to 5 ppm in 2012, which might require a negotiated waiver to maintain operations at current performance levels. However, an Air Quality District regulation states that boilers operated at less than 10% of their annualized maximum power are exempt from this emission requirement, and the maximum annualized power usage for the ARC boiler has been 7%. This implies that ARC will have no problem with this regulation unless their current capacity increases by more than 30%. If the Air Quality District changes this exemption or if a negotiated waiver cannot be obtained for the expected lower NO_x level limit, the ARC arc jet complex will have to cease operations. Although the ARC boiler is old, it was inspected in 2009 and found to be in sufficient condition to operate for an additional ten years.

- 150 MW Power Supply. Health monitoring of the power supply has detected initial component breakdown for these long-lead items.

- IHF Flow Controls, Test Chamber, Diffuser, and Gate Valve. Increasing heating and flow rates, non-optimized condition and configuration feedback, and longer run durations are causing failures in undersized components.

- AHF and PTF Diffusers and Heat Exchangers. To address vacuum and water leaks from corroded and fatigued components and seismic regulations.

- SVS Plenum, Piping, and Tanks Seismic Upgrades. To address potential personnel injury, vacuum and water leaks, and structural failures from aged plenums, pipes, and water tanks during earthquakes and seismic regulations.

5.3.2 Johnson Space Center

The current age and estimated life expectancy and remaining service for the subsystems that support the ARMSEF are shown in Table 5.5.

Table 5.5. JSC Arc Jet Complex Subsystem Summary
[Riccio, JSC Experimental Heat Transfer and Upgrade Status]

Subsystem	Component	Life Expectancy	Current Age	Remaining Service
Steam Generation	Boiler	20	20	0
Heat Rejection	Cooling Tower	15	1	14
Heat Rejection	Pond	20	2	18
Arc Heater Coolant	Pump	20	0	20
Test Gas Storage & Delivery	Digital Valves	20	20	0
Test Gas Storage & Delivery	LN2	20	30	-10
Test Gas Storage & Delivery	GN2, GO2	20	2	18
Data Acquisition & Control	Modcomp	15	30	-15
Data Acquisition & Control	ABB/ASEA	15	6	9
Model Control	Insertion	20	30	-10
Power Conditioning	Load Bank	30	5	25
Power Conditioning	Transformer	30	1	29
Power Conditioning	Rectifier	30	43	-13
Vacuum	Ejector	20	20	0

Several systems and components in the summary are at or beyond their nominal service life. This is typical for Government technical facilities; these systems and components are not considered to be in critical condition. This analysis was endorsed by JSC in the March 18, 2010, presentation to the AJEWG.

JSC tracks unplanned downtime, and this measure can provide a basis from which facility condition can be inferred. In 2006, 2007, 2008, and 2009, JSC experienced approximately 11 weeks, 0 weeks, 4 weeks, and 0 weeks, respectively, of unplanned downtime. This equates to an average availability of 90% over the four-year period.

JSC has developed a plan to revitalize the ARMSEF, and systems upgrades and estimated costs are shown in Table 5.6.

Table 5.6. JSC Arc Jet Complex System Upgrades
[Riccio, JSC Experimental Heat Transfer and Upgrade Status]

ARMSEF Upgrades	Purpose/Improvement Area	Cost ($K)
Steam Ejector Tune-up	Reliability & Performance	350
Parallel Digital Valves	Reliability & Performance	200
Data System Modernization	Reliability	500
Test Position 2 Multi Arm	Model Throughput	250
High Capacity Steam Ejector	New Capability, Reliability	1800
Test Position 1 Real Time Channel View	New Capability	100
Test Position 3 Design & Fabrication	New Capability, Model Throughput	1000
Mars CO2 Test Gas System	New Capability	800
Rectifier Component Modernization	Improved Capability, Reliability	1000
Test Position 1 Plug & Play Models	New Capability, Model Throughput	250
Coolant System Redundant Pump Leg	Reliability	500
Total ARMSEF Upgrades		**6750**

JSC's investment strategy for the ARMSEF also identifies several areas of concern, and these are as follows:

- Vacuum and Ejector System. Many components are aged and this system is in need of complete overhaul. The redesign will include increased capability to permit a higher mass flow rate.
- Boiler System. The boiler causes frequent outages and downtime for the facility, and the harsh operating cycles lead to an ongoing risk. A direct boiler replacement has been designed, and a Pre-Phase A design for an alternate system to provide steam from Building 24 has been completed. Although the JSC steam boiler is currently operating at the end of its 20-year design life, the boiler was inspected and certified in January 2010 and is in sufficient condition to operate for another 7 to 10 years.
- Diffuser System. The diffuser system is aged, and a failure puts the facility at risk for long downtime. A replacement is required, but is difficult to implement without an extended downtime.

5.3.3 Air Force Arnold Engineering Development Center

The Air Force developed an investment strategy to address revitalization needs and to increase technical capability. These investments will close the current gap between AEDC and NASA arc jet capability. Included in this plan are the following:

- H2/H3 Stilling and Mixing Chamber. Design and fabricate a stilling and mixing chamber and housing compatible with both the H2 and H3 heater stacks to provide additional total temperature control at lower enthalpy with mixing air, and to provide improved flow enthalpy profiles.
- H2 Diffuser Replacement. Replace the existing 1960s-vintage diffuser to provide higher thermal efficiency and capacity.
- H2 Model Positioner System (MPS) Axial Controls Upgrade. Replace the 1980s-vintage MPS axial controls.
- Mid-Pressure Arc Heater. Replace the existing Huels heater in the H2 arc tunnel with a high-performance, segmented arc heater utilizing the demonstrated technology of the operational H3 heater to provide improved simulations for many DOD hypersonics test points, with higher enthalpy, efficiency, and cleaner flow. Upgrades to existing facility power supply and cooling water systems will be required to support the higher-performance, segmented arc heater for longer run times. As shown in Figure 5.2 below, the implementation of this investment (the light blue dotted area) will expand AEDC's arc jet performance to fill much of the pressure/enthalpy gap between the Air Force's and NASA's current capabilities.

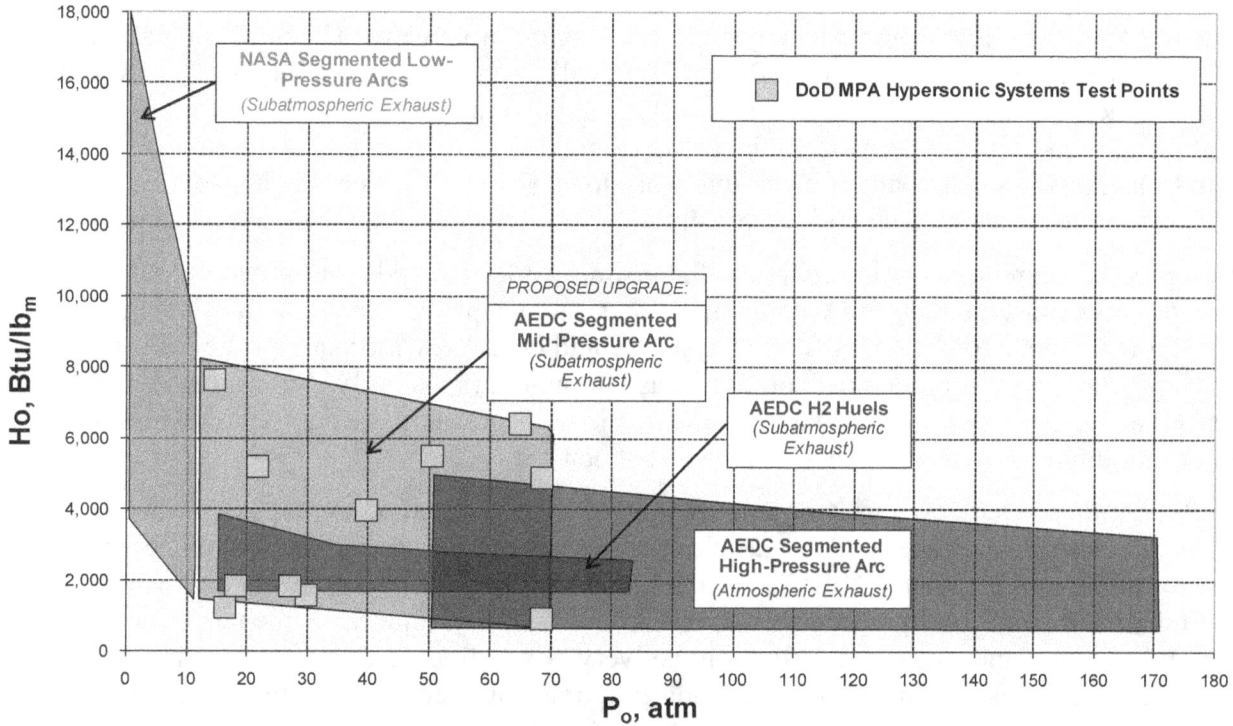

Figure 5.2. AEDC proposed upgrade to mid-pressure arc heater
[Smith, AEDC Arc Jet Facility Capabilities]

5.4 Mothballing the Existing Arc Jet Complexes at ARC and JSC

Discussions with arc jet operators at ARC, JSC, and LCAT made it clear that mothballing a facility for more than three years effectively equates to closing that facility. Costly infrastructure, especially boilers and steam vacuum system, deteriorates rapidly when it is not operated.

Mothballing generally removes 85% of the annual cost to operate and maintain a facility, but requires an initial investment for the basic preservation of the systems and equipment that might later be reactivated. Mothballing requires an annual investment to provide safety and fire protection for the complex and a minimal level of equipment and system maintenance and operation. If a mothballed facility is reactivated, a subsequent investment in maintenance, repair, revitalization, and staff training will also be required before the re-energized equipment and systems can become recertified for operation.

Mothballing of both complexes for more than 3 years is essentially the same as closing the facilities. The estimated cost for mothballing the existing arc jet complexes over a 30-year period is located in a table of cost comparisons in Appendix G: Investment Options, although it is unlikely that NASA would continue to mothball a facility for that long. JSC evaluated the possibility of mothballing and reactivating their arc jet complex, and their estimates are several million to mothball and several million to reactivate. They also estimated that reactivation will take 18 to 24 months and will be driven by the loss and replacement of critical skills and the training required.

5.5 Throughput Efficiencies

Existing NASA arc jet throughput can be increased in multiple ways. The implementation of the investment strategies developed by ARC and JSC will improve the reliability and the availability of the tunnels and supporting systems and will enable the tunnels to be test ready more often. Adding a second shift at the ARC arc jet complex and utilizing extended shifts and overtime at both Centers would help manage a variable work load. (Note: This is not being advocated by the AJEWG, but is considered an option to maintain testing schedules and increase capacity.)

An option that can increase throughput with existing workforce and work schedules is to replace a single-model insertion system with an insertion system capable of holding multiple models so that, for example, a five-model insertion system could hold one calibration probe and four test articles. This type of arrangement could increase throughput by simply increasing the number of items that can be cycled through the test stream and by decreasing the number of calibration probes required for the test articles and the test stream.

Another option to increase throughput is via a more efficient method to prepare, mount, and check out calibration probes and test articles before they are introduced into the tunnel test bay. The installation of a "plug and play" system that has a model/probe assembly room with instrumentation and cooling water system checkout capability adjacent to the test bays will improve the movement of test articles from delivery box to final test data. A significant improvement to this option is through the addition of a chamber or room attached to the test chamber where a second or third assembly stand of multiple-model insertion systems could be ready to "plug in" to replace the assembly coming out of the chamber. This chamber/room could be pumped down to match the test chamber pressure for remote insertion and removal of the test model assembly or vented to atmospheric pressure to facilitate model changes while the arc jet continues to operate. The adjacent chamber/room could contain two or three multiple-model assemblies that would allow many repeat runs at a set condition or allow minor changes in test conditions without de-energizing the arc jet. This would minimize or eliminate the recalibration that is often required when trying to match a previous set point or when interpolating data between runs that had similar set points, and would allow many more test models to be run for any given arc jet operation.

An additional area of improvement is redesigning the arc heaters for maintainability to provide a more stable arc operation, reduce the amount of time and effort required to replace electrodes, and to allow the easier diagnosis and repair of water leaks.

All of these options will improve the throughput efficiency. However, NASA should carefully assess future testing requirements and the costs associated with meeting these requirements before implementing any improvements designed to increase throughput.

5.6 Summary

Finding 5.1: Facilities need to retain operational status, as mothballing an arc jet complex beyond 3 years is tantamount to closure because of the degraded nature of the supporting infrastructure.

Finding 5.2: There are opportunities for improving capacity at both complexes.

Finding 5.3: Staffing levels at both NASA complexes appear to be high.

Conclusion 5.1: Experienced personnel with specialized arc jet training and skills are valuable and irreplaceable. Arc jet facilities must maintain a skilled workforce for safe and reliable operation.

Conclusion 5.2: A more thoughtful approach to building new arc jet capability should consider the larger, long-term Agency mission requirements, as presented in Section 3.0 of this report; integrate the full institutional cost, capability, sustainment, and life cycle into Agency strategic planning; and consider the difference between what is known about arc jet technology and the knowledge needed to advance the state of the art to design an arc jet facility for future capability.

Conclusion 5.3: In an effort to reduce costs, an independent assessment of roles and responsibilities at both the Ames and Johnson complexes should be completed and staffing levels reduced by up to 30%.

6.0 Conclusions and Plan Forward for NASA Investment in Arc Jet Facilities

NASA needs to invest in an arc jet complex that can support certification of missions that deliver crew to the Martian surface and return them safely to Earth, as well as sample return missions from beyond LEO. These missions remain the long-term centerpiece of NASA's exploration vision. The AJEWG considered numerous possible options for NASA's investment, which are discussed in Appendix G: Investment Options. The AJEWG concluded that the required test conditions are beyond the capability of any existing facility, so new construction is required. In fact, the required combination of convective and radiative heating rate is beyond current state of the art, so technology development is required for the arc heaters themselves. Although the details of the heaters are not defined, there is enough information to define the infrastructure required to support the tests, and to begin to design it.

6.1 Required Capability

The AJEWG asked the user community to identify features that should be included in a new arc jet complex. Their responses are summarized in Table 6.1. The test specifications for safe return from Mars indicate that 3,000 W/cm^2 of convective heating and 1,400 W/cm^2 of radiative heating applied to a 6-inch stagnation model at 40 kPa pressure would provide excellent simulation of anticipated conditions. This is a reasonable starting point for sizing facility infrastructure. Heating and shear levels that can be applied to wedges and panels of similar scale should also be considered. Earth return of small science payloads at very high velocity and entry to Jupiter require very high heating rates, both convective and radiative, but relatively small models are acceptable and larger deviations from nominal conditions may be accepted for robotic missions. If the arc jet complex infrastructure is sized for crewed Mars return, science missions to other destinations will need to test in non-optimal environments, but they will be much more relevant than anything that can be done in existing facilities.

Arc heaters that can deliver these heating levels at these scales do not exist today. A technology program is required to develop new capabilities. Important elements of technology development are discussed Section 6.3. Even without knowing the details of the arc heaters that will need to be housed in the complex, much of the infrastructure design can begin immediately and complex construction can begin before the new heaters are fully designed.

In 2007, an AOA study was conducted by Ares Corporation to evaluate options for arc jet support for certification of Orion at Lunar return conditions for the LEAF project. A similar study is recommended for Mars Direct Return certification requirements. Since certification requirements for this mission are not yet defined, a parallel effort should be conducted to lay out a certification and mission assurance roadmap for all missions to be supported by arc jet testing. Elements of the proposed roadmap are discussed briefly in Section 6.4. Even before a comprehensive study is performed, we can state that a power supply of at least 100 MW will be needed, which is substantially larger than the supplies at existing complexes.

A single arc facility will not support all mission types. Arc jet complex infrastructure should have multiple bays to accommodate different arc heaters. There should be at least one bay dedicated to arc test technology development and low-TRL material evaluation, so that these activities can be continuously pursued in parallel with mission support testing. A minimum of four bays will be required, and six bays would be preferred.

Evaluation of Arc Jet Facilities Report

Table 6.1 Future Arc Jet Needs: Summary of Responses from End Users

	Test Conditions	Model Size	Test Gases	Combined Convective Radiative Heating	Control of Flow Conditions	Knowledge of Test Conditions	Instrumentation	Cost, Access, Throughput	Operations
Commercial LEO	Thermally similar to current facilities	~ 10 cm dia x 40 cm long						Cost and access wise, improved access and cost structure to ease barriers to testing and material development iteration	Test prep and instrumentation area and check-out independent of the facility to facilitate quick, accurate and efficient testing and use of the facility.
ISPT (Munk)	At least the existing capability (PTF through IHF)	Increased sample size is a nice-to-have; significant scale-up is not necessary (but we may have come to that conclusion based on facility limitations)	Alternate gases	Combined environments	Improve quality (repeatability of test conditions)	Improve test condition knowledge, especially shear	Improve instrumentation	Improve cost and throughput	
Orion (Durrant)	Flexibility and control of heat rate, enthalpy, pressure, shear and boundary layer thickness	Larger model size (20" x 20") panel for OML features). Adequate accommodation of curved panels.		Yes	More automated test condition settings	Better flow monitoring capabilities	Improved instrumentation	Improved test throughput: improved model installation approaches, more models per run	
Don Curry	Restore capability for lunar and planetary missions				Improve uniformity, reduce contamination				
Orion (Bouslog)	Expanded test envelop	Panel models 18" x 18" to 30" x 30"; Stagnation models 4" - 10" diameter; Wedge models 5" to 12" square	New gas mixtures	Run temperature and pressure profiles simultaneously. Radiant heater zones for	Match time-dependent heating profile for mission trajectory	Different arc heater designs to compare differences in nominally identical conditions.			Facilitate nozzle changes, plenum changes, arc heater modifications

Evaluation of Arc Jet Facilities Report

	Test Conditions	Model Size	Test Gases	Combined Convective Radiative Heating	Control of Flow Conditions	Knowledge of Test Conditions	Instrumentation	Cost, Access, Throughput	Operations
Mars Entry (Steltzner)	0-3,000 W/cm2 convective + 1-3,000 W/cm² radiative, 1-1,000 Pa shear, 0-20 atm		CO_2 "nice to have"	0-3000 W/cm² convective + 1-3000 W/cm2 radiative		Heat flux and pressure probes, flow chemistry. thermal gradient effects.			
HIADS (Cheatwood)					More uniform freestream (PTF has swirl)	Better characterization of freestream		Increase throughput	Better adherence to facility schedule via more reliable facilities
					Ability to dial in flux and surface pressure to match desired profile				
Gas Giants (Venkatapathy)	Up to 20,000 W/cm² convective, 40,000 W/cm² radiative (Jupiter). Up to 8,000 W/cm² convective, 5,000 W/cm² radiative other gas giants	2-4" stagnation; turbulent flow at moderate conditions on large samples	Air, CO_2, H_2/He	10,000 W/cm² convective, 10,000 W/cm² radiative	Turbulent flow	Improved knowledge of test environment		If the cost/run formula is modified, we will see increased use and reduced risk (or better mass)	
ETDP - Wright	Earth return: 1,500 W/cm² convective, 2,200 W/cm² radiative, 40 kPa pressure, 450 Pa shear, 100 MJ/kg enthalpy	Turbulent flow on larger models at flight-relevant conditions. Models >6" with high heat rate, pressure, shear	CO_2, Air	Yes	Trajectory-based testing	Integrated analysis/testing services to maximize understanding of results		Cost/sample no more than current, throughput at least matching current complexes	Clean separation between research and production facilities (R&D needs currently sacrificed for sustaining engineering testing)

Support for different test gases is needed, although individual bays do not need to be plumbed for all gases. Pure air, pure nitrogen, and mixtures of oxygen, nitrogen and argon should all be provided. Large-scale testing in CO_2 at moderate combined heating rates and small-scale testing in H2/He at very high heating rates should be supported.

Several respondents stressed the importance of test profiles that mimic the variation of heating and pressure throughout an entry trajectory. The complex infrastructure should support variation of power and pressure throughout a test entry.

Beyond the requirements for mission-relevant test conditions, responses in Table 6.1 consistently note the importance of having better knowledge of the conditions to which the model is exposed. New test boxes must be designed with adequate optical access for flow diagnostics and particle identification. They should also support thermal and pressure surveys of the core flow.

Respondents further noted that cost per test-article exposure is an important metric. Several indicated they would perform more tests if throughput were increased. In recent years, 60% of exposures at ARC and 40% of exposures at JSC have been calibration probes. The most effective way to improve throughput is to increase the number of stings. The proposed design for LEAF included six model arms, with two additional arms for calorimeter and pitot probe. A similar configuration is appropriate for new tunnels.

6.2 Options for New Capability

Options for providing upgraded capability include:

- Add a new arc jet facility to an existing complex
- Build a new arc jet complex with one arc jet facility to augment existing capability
- Build a new arc jet complex with multiple arc jet facilities to augment and replace existing capability

The LEAF AoA compared the first two of these options. It concluded that the preferred option for delivering Lunar return capability was the first option, to add a new arc heater to an existing complex (ARC's), but noted in passing that construction of a new complex would be required for Mars return capability. That conclusion should be reviewed in a new study focused on Mars return requirements, but it is driven primarily by the combination of model size and heating rate requirements: existing arc tunnels cannot deliver sufficiently heated gas to models of the required scale. Furthermore, some compromises related to power supply, test box layout, diffuser, and steam generation were made to accommodate LEAF at ARC. Schedule was a driver, because an extra year of construction for a new complex did not permit adequate ground testing prior to Orion test flights. With more schedule latitude, such compromises should be revisited in a new study.

When new construction is evaluated, several sites should be considered as viable options. The location should be chosen to ensure ready access to plentiful and cheap power and water, low impact on air and water quality, convenient access for qualified personnel and test customers, and minimal intrusion of operations into the surrounding communities. A site that has other major test complexes provides greatest flexibility for sharing support personnel when test demand is low, thereby mitigating annual operating costs. It is expected that NASA would consider both ARC and JSC, but other sites, such as AEDC, might be considered if they clearly offer advantages in these criteria. Although existing arc jets at AEDC do not support NASA mission profiles, the Air Force is interested in extending their capability into a range of conditions that more closely approach NASA needs. However, sharing infrastructure would require collaboration across agencies for strategic planning and management and to ensure adequate accessibility.

New construction provides the best opportunity to deliver state-of-the-art technical capability with efficient infrastructure and operations. These efficiencies can deliver a life-cycle value that is competitive with the upgrade of existing facilities. New construction will also allow NASA to eliminate the deferred maintenance accumulated in existing complexes and will help reduce the overall average facility age and improve overall facility condition.

The estimate of the 30-year cost for new construction of an arc jet complex comprises the initial investment, the annual operations and maintenance, and the periodic refurbishment necessary to keep a complex viable and recapitalized. Two cost estimates were developed for the initial investment to build a new construction arc jet complex that included the LEAF capability and three other tunnels that would duplicate the existing capability at ARC. The first was developed by the ARES Corporation and was included in the original LEAF AoA. The second estimate was for the same capability identified in the AoA but was performed by facilities engineering personnel at ARC. The estimates were based on a narrow focus on Orion requirements, however, which did not consider the cost of enhancements that could address requirements from other missions. A tunnel to support higher-energy requirements would cost more. Since the requirements for a new tunnel have yet to be determined, the 30-year cost estimate is provided as a range of the costs that would likely be involved. The initial investments identified are the basis for this estimate, but are considered to be the lower bounds of the range, and a 50% premium on the escalated ARC estimate is used to determine a reasonable higher bound.

Regional differences in construction costs would also be a factor in the estimate for the initial investment required, and these differences are captured in Table 6.2 below. This data is from the *Engineering New-Record,* 20-City National Index, and regional differences for several cities are captured in the Construction Cost Index (CCI) and then normalized to construction in San Francisco (the CCI Factor). The CCI uses local prices for general construction materials, such as concrete, lumber, and structural steel and local union wages. However, it does not account for specialized equipment, such as boilers, steam ejectors, and high-power electrical distribution components, since the cost of this type of equipment does not have a regional bias and differences in installed cost are primarily due to transportation factors. From the CCI Factor column, general construction materials and labor costs in Dallas and Birmingham (AEDC) are 40% lower than in San Francisco and Cleveland. This 40% difference will be reduced when high-cost, specialized equipment is included in a construction cost estimate, and a reasonable difference for total construction costs is 25%.

Table 6.2. Regional Construction Cost Index

City	CCI	CCI Factor
Baltimore	6011	0.62
Birmingham	5720	0.59
Cleveland	9989	1.03
Dallas	5339	0.55
Los Angeles	9770	1.00
New Orleans	4935	0.51
San Francisco	9728	1.00

Data from *Engineering News-Record*, 20-City National Index
Construction Cost Index (CCI) as of March 2010
CCI Factor relative to San Francisco

The second component of the 30-year estimate is the annual cost for operations and maintenance, and there are also regional differences for labor costs. Half of the current ARC arc jet complex annual operation and maintenance costs is labor (12% for civil servants and 37% for contractors). This annual number provides a basis for the cost of operations and maintenance in a new complex, since the upstream and downstream infrastructure will still be shared, and the simultaneous operation of multiple tunnels is unlikely. A range is used for a new complex at ARC to reflect the implementation of Conclusion 5.3 and to allow for the possibility the new complex will require more resources to operate and maintain, and this range is escalated at 2% per year for inflation. The cost of labor will also reflect regional differences, and two data sources exist to determine regional discounts for NASA locations. The first is the data NASA accumulates for its workforce and reports on nasapeople.nasa.gov. Average annual salary data can be found by linking through The NASA Workforce Profile to Workforce Measures, and average salaries at ARC are 12% higher than at JSC. The second source is presentations provided during the Center visits. Each presentation contained current contractor labor costs, and a comparison of these costs reveals that the average salary at ARC is 4% higher. A reasonable estimate of the regional discount at JSC for the combined workforce is 5%. This regional difference is more pronounced for AEDC. The AJEWG contacted the AEDC procurement office to identify the labor rate associated with the Aerospace Testing Alliance contract the Air Force uses to operate and maintain their wind tunnels and testing capability. The Air Force pays an average contractor labor rate that is 22% lower than NASA pays at the ARC arc jet complex. A reasonable estimate of the regional discount for a workforce at AEDC is 20%.

The last component of the 30-year estimate is the cost to revitalize and recapitalize the complex. The facility presentation from ARC includes the projects accomplished since 1990 and an investment strategy intended to refurbish, upgrade, or replace capability (revitalization) and infrastructure (recapitalization). These items form the basis for the ARC revitalization and recapitalization estimate. Accounting for procurement of specialized equipment and for the labor required for installation and activation, the effective regional discount at JSC and AEDC for revitalization and recapitalization is estimated to be 20%. Such a discount would probably not be fully realized in a new build, but it points to potentially significant cost differences by location.

All of these estimated costs are captured in Table 6.3 [deleted in public version of report] and are graphed in Figure 6.1. These data indicate regional labor differences are worthy of consideration in the selection of location for a new complex.

6.3 Considerations for Upgrading to Future Capability

The first option in this section, add a new arc jet facility to an existing complex, assumed utilizing the existing complex as is. While building augmented capability that takes advantage of existing infrastructure was reasonable to consider at the ARC arc jet complex, the AJEWG did not consider this option to be technically reasonable at JSC. Making ARC healthy would involve refurbishing and repairing the arc jet systems and components. To bring JSC to an equivalent technical capability to ARC's would require replacing most of the complex, including the arc heaters, water systems, vacuum systems, and power supply and adding tunnels. It would be possible to make JSC's existing complex healthy, but upgrading the capability would be equivalent to a new build. So for JSC, the equivalent to "Make ARC healthy, build a new tunnel at ARC" would be "Make JSC healthy, build a new complex at JSC." There were no existing estimates to use for this option, so it was extrapolated from the data available to the AJEWG.

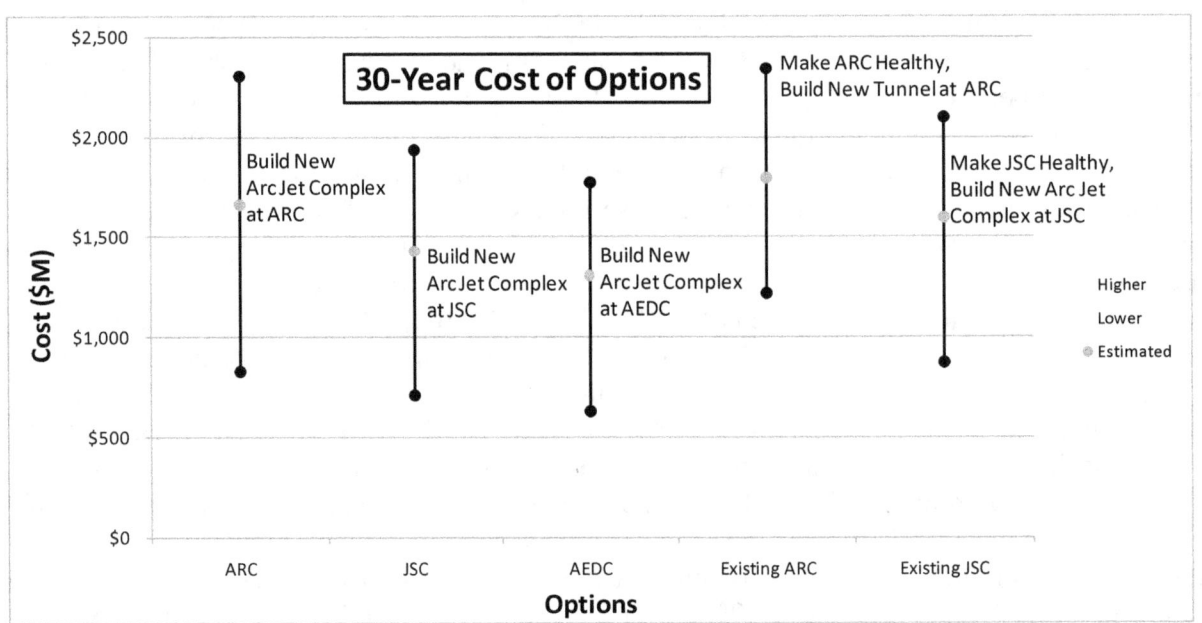

Figure 6.1. Estimated 30-year cost of options

The cost analysis predicts that both options where existing NASA arc jet complexes are made healthy and upgraded with new capability to meet the 30-year future requirements would be more expensive than building a new complex at that Center, and neither is considered a good investment strategy.

The estimates of the 30-year costs of the options indicate that trying to take advantage of existing capability will be the most expensive of the options. Also, over the 30-year period of interest, investment differences between locations for new construction are significant enough to be a factor in the location decision.

6.4 Technology Development Requirements

To provide the performance capability needed for a future arc jet complex, a significant investment in arc heater technology will be required. There are sizable gaps between the current state of the art in arc heater performance and that required for missions beyond LEO. Also, arc technology is needed to provide a better understanding of the arc jet flow field. Some of the issues and possible ways to address them are presented below.

6.3.1 Understanding the Arc Jet Flow Field

There have been long-standing differences in the results obtained from testing the same materials in the ARC and the JSC arc jet facilities. It is important to know which facility produces the flow field that more closely approximates the true entry environment and to understand the source of the differences. Possible causes for the difference in material response are the incomplete mixing of O_2 and N_2 at JSC, the addition of argon at ARC, or the difference in enthalpy and pressure profiles between the two facilities. To resolve this issue, a number of activities are proposed:

1. Perform a "round robin" series of tests at ARC, JSC, and Boeing LCAT using the same test material, flow-field conditions, and configuration. This will permit a comparison of the effects of simulated air (JSC), air with a small percentage of argon (ARC), and pure air (LCAT).
2. After the test series above, JSC should reconfigure either their TP-1 or TP-2 arc heater by adding a ring electrode at the cathode and operate this heater first with pure air and then on air with argon added in the same ratio as that used at ARC, and compare the results.
3. Compare arc heater configurations, enthalpy and pressure profiles, and test conditions (pressure, heat flux, and Mach number) between the two facilities.

For simulation of entry into the atmosphere of other planets, arc jet operation will be required with other test gases such as CO_2, He/H_2, and N_2. Again, the TP-1 or TP-2 arc heater at JSC would be ideal for this development to learn how to operate with these test gases and to obtain arc reliability and performance data. JSC personnel should draw upon the ARC experience with He/H_2 from the previous operation of the Giant Planet Facility and upon Marshall Space Flight Center's ongoing experience with a 1 MW He/H_2 heater.

Development of improved flow-field instrumentation is critical to the understanding of material test results. This includes both intrusive and nonintrusive measurements. Nonintrusive diagnostics are particularly important because the flow field is left undisturbed. The challenge has always been to obtain meaningful measurements in an unsteady, high temperature, luminous, ionized flow produced by arc heaters. The development and use of these advanced instruments is critical to a better understanding of the arc jet flow.

6.3.2 Closing the Simulation Gap for Future Missions

For missions beyond LEO, the gap is significant between present day arc jet facility capabilities and those required for future missions (see Section 4). Major advances beyond current state of the art in arc heater performance and facility capability are needed. Several avenues for arc facility technology and development are suggested below:

1. **Arc Heater Performance Improvement**: The segmented arc heater design is considered mature. It has been incrementally improved over the past 50 years in both performance (enthalpy) and reliability. However, additional minor improvements are still possible and

should be pursued. Development of improved cooling techniques for the electrodes and segments will allow hardware survival at higher wall heat flux and higher arc current, which will produce higher enthalpy. However, no more than a 10% to 20% increase in performance can be expected from improved component cooling.

2. **Arc Jet Flow Field Augmentation**:
   a. Many of the missions to Mars and beyond will experience significant radiative heating, in addition to the high convective heating that arc jets can provide. Augmentation of the flow field with an external radiation source will be required. Arc lamps with beams focused on the test model surface provide one possible method for radiant heating. Laser beams provide an alternative way to provide substantial radiative heating, but the effect of a single wave length source would need to be investigated and understood. ARC personnel are already studying radiation augmentation methods as part of the planning process for the LEAF arc heater.
   b. Another method of adding energy to the flow is through magneto hydrodynamics (MHD). This method has the advantage of accelerating the flow velocity directly in the supersonic region of the nozzle without adding significant heat. In principle, the MHD method is sound. In the past, significant hardware problems (wall arcing) and non-symmetric flow field acceleration have resulted in limited operational success.

3. **Non-Standard Test Techniques:** One test technique that has been demonstrated is to allow the arc discharge to pass through the nozzle and attach downstream, either to the downstream face of the nozzle (ARC concept used on the Giant Planet Facility) or at the base of the model sting (Boeing concept). This technique can increase both the convective and radiative heating simultaneously, since the arc temperature is 10,000 to 12,000 degrees Kelvin and the arc is either near or surrounding the model. This technique may be suitable only for light gases, but should be studied for air and other gases as well, because of its potential for providing significantly higher heat flux on the test model.

4. **Non-Standard Arc Heater Configurations:** One method used in the past to build larger (higher power) arc heated facilities was to manifold the flow from several arc heaters together. In one concept, either four or five arc heaters are located at 90 degrees to each other with the flow from each combined and discharged through a single nozzle. Another arc heater configuration has incorporated an anode on one end and a cathode on the other end in a transverse direction to the flow exiting from the nozzle in the center. LaRC previously developed this type of "double-ended" arc heater. However, while these devices provide a means to scale to larger sizes, they do not produce higher performance (enthalpy).

6.3.3 Sustained Arc Technology Program Required

To better understand the arc jet flow field and to provide the enabling technologies to narrow the gap in arc facility performance, a dedicated arc jet development effort is required. NASA should provide continuous funding over the next 5-10 years to advance the arc jet simulation capability and develop the enhanced performance necessary to produce the next generation of arc heaters.

ARC and JSC should jointly manage the technology development effort with a common purpose and goal. The actual development tests can be performed at either or both complexes to maximize the progress toward improving arc jet flow field performance and understanding.

6.5 Certification and Mission Assurance Roadmap

The level of ground testing for proposed missions depends on NASA's risk posture, commitment to flight testing, and the level of confidence in numerical simulation of the entry physics and material chemistry. In the past, these issues have been treated differently by individual missions and have sometimes been addressed only qualitatively. Recent programs, including Orion and COTS, have specified mission reliability requirements, and allocations have flowed down to TPS. Verifying compliance with these allocations is proving to be extremely challenging. If planetary protection requirements for return to Earth from Mars continue to impose 1:1,000,000 failure probability for entry, descent, and landing, as specified for Mars Sample Return, it is not clear that even the most rigorous ground test campaign can retire the mission risk.

NASA needs urgently to review certification requirements and mission reliability allocations for TPS. To meet the technology development schedule, this effort should be completed in the next year and integrated into the capability requirements to ensure they are consistent with certification needs. Recent work on certification strategy for Orion and current work on human rating requirements for commercial crew transportation should be generalized into a common framework for assuring TPS reliability for all missions. Required levels of reliability may differ among missions, and choice of verification approach may also vary, but the methodology should be consistent. The certification approach should address the contributions of analysis, ground test, and flight test toward understanding material and subsystem behavior.

Several respondents to the AJEWG survey on arc jet needs stressed the value of flight data that provides traceability for ground test and analysis. The optimal arc jet complex must be defined in the context of the complete TPS development and design process. As an example, the cost of the proposed LEAF facility is about 20% of the estimated cost of a single subscale flight test dedicated to retiring TPS risks for Orion (LEX flight test project) [McDonald, LEX Study Team Report]). It might not replace the flight test, but the flight test project manager indicated that it would help him to sign the Certificate of Flight Test Readiness and it would augment the value of the flight test by providing a much larger data set for comparison and correlation. Technology programs can balance investment among ground facilities, analysis methodology improvements, and engineering instrumentation on missions and dedicated flight tests.

Optimal use of an arc jet complex must include low-TRL development work with direct support of missions. Much of the effort required to qualify materials can be managed through material and subsystem development programs, rather than being borne by individual missions. NASA should develop a family of qualified materials for a range of entry environments, to be adopted and carried to certification by individual missions.

6.6 Transition and Institutional Management

The skills and knowledge of the technicians, engineers, and managers at existing arc jet complexes will be needed to develop the next level of arc jet capability and, subsequently, to operate the new complex. Provision will need to be made to maintain safe operations while drawing down or transferring workforce to support the new build. A break in continuity, which could result from the immediate threats to maintenance and operation (M&O) funding for

Johnson, could create an unnecessary loss in personnel that would compromise the introduction of new arc jet capability.

Existing infrastructure and arc heaters must also be managed efficiently through the period of transition. The agency has benefited from strong cooperation between the two NASA complexes in recent years. This positive relationship should be encouraged and supported. The arc jet managers at ARC and JSC proposed to the AJEWG that their complexes should be managed cooperatively as one Agency portfolio. They indicated that the requirement for individual facility robustness could be relaxed by leveraging the capacity of the two complexes. They suggested deferring all proposed refurbishments until calendar year 2016. This proposal fits well with the AJEWG conclusions and timeline describing the transition of the facilities' technology in the following section. The AJEWG supports this creative solution to stabilizing near-term arc jet operations.

Common management of agency arc jets will apply a common cost basis for testing. Until now, separate complexes have been funded by different organizations within NASA, so they have had distinct cost recovery practices. Furthermore, customers collocated with the facilities have benefited from convenient access and low-cost test opportunities. More equitable access to the test complex for all customers could engage a larger community and would foster additional innovative TPS solutions. Facility operators and customers all recognize that full-cost recovery decreases utilization and stifles low-TRL material development. On the other hand, suggestions that users should not be responsible for any of the facility cost would reduce incentive to rigorously plan and execute their test programs. NASA needs to generate a cost recovery structure that maximizes the strategic benefit of the arc jets.

NASA's funding and governance of this critical institutional technical capability has been inadequate. The current list of deferred maintenance provides evidence that the lack of reliable and sustained funding has contributed to deterioration of arc jet systems. Revitalization activities have often been funded by individual projects or missions, so that capability has been focused on specific needs rather than being optimized for maximum value to the agency. To ensure sustainable and efficient operation of a new arc jet capability, NASA needs to commit to continuous and consistent institutionalized funding and resource management.

6.7 Development Schedule

The AJEWG concludes that the Agency should immediately begin efforts leading to the construction of a new arc jet complex, and the notional schedule shown in Table 6.4 identifies the top-level activities that will lead to this new complex becoming operational. The requirements definition necessary for the infrastructure design should begin immediately. The project milestones include a PDR in 12 months, a CDR 12 months later, construction groundbreaking 12 months beyond that, and the new complex becoming operational 5 years from now.

In parallel with the infrastructure build is the technology development required for a new arc jet complex. This technology development will focus on the radiative capability associated with high-energy entry conditions and how this level of external radiative heating will be generated and focused on a test model; extend the state of the art in arc jet heater operation at continuous, high power levels through improved cooling methods and nonstandard configurations and test techniques; and improve the flow field capability required for condition validation, data acquisition, and test sample certification through alternate cathode configurations, alternate test gases and gas combinations, intrusive and non-intrusive instrumentation, and increasing enthalpy through the use of magneto hydrodynamics.

Also shown in Table 6.4 is the transition of the existing ARC and JSC arc jet complexes through the infrastructure build. The last Shuttle program launch, STS-133, is scheduled for September 2010, and the JSC complex should transition to one-shift operations immediately after this launch. JSC should also transition into support for the technology development required for the new complex, and should plan for an orderly phase-out after the completion of the technology development efforts. ARC should reduce their operations by December 2010 to a capacity required by the Agency. Operations at the ARC complex should continue through construction of the new complex to provide a capability to address issues associated with integrated systems testing, but the phase-out of the ARC complex should coincide with the new arc jet complex becoming operational.

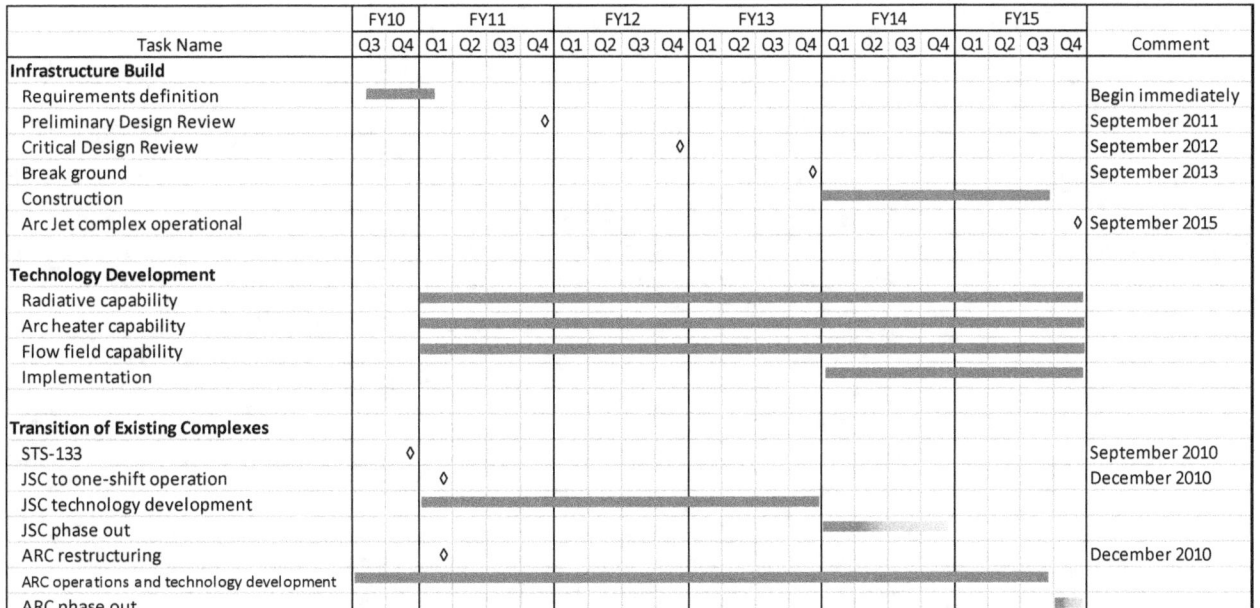

Table 6.4. Notional Schedule for Transition to New Arc Jet Capability

6.8 Summary

Finding 6.1: The return on investment of rehabilitating and upgrading existing complexes, or building a new complex, would be positive. Although arc jet test cost is significant, it is a necessary cost for achieving many of NASA's proposed missions. The total cost of construction is on the same order as one or two subscale flight tests.

Finding 6.2: Arc jet users are cost sensitive, especially commercial customers. They will do more testing if the cost per test is reduced.

Finding 6.3: NASA must have skilled and knowledgeable personnel to develop and operate the next generation of arc heaters.

Finding 6.4: NASA funding and governance of critical institutional technical capability is inadequate. Arc jets have had intermittent funding over several decades, and preventative maintenance has suffered. Strategic Capabilities Assets Program (SCAP) funding to Ames

Research Center has allowed them to work down a backlog of deferred maintenance, resulting in improved availability. JSC has benefited from Shuttle and Constellation program funding, which is expected to go away in 2010.

Conclusion 6.1: NASA needs to make a serious investment in arc jet infrastructure. Investment in the range of $200-500 million in addition to continuing operating costs will be needed in the next 10 years. Resource requirements will be diligently challenged and will limit future technical options.

Conclusion 6.2: Decisions on transition to a new complex need to consider retention and sustainment of highly-qualified staff. Recent cooperation between ARC and JSC has had positive benefits for the Agency and that relationship should be fostered and supported during the transition.

Conclusion 6.3: Cooperative management of JSC and ARC complexes is recommended while a new complex is being developed.

7.0 Findings and Conclusions

Finding 2.1: Arc jet testing has proven to be a core competency and required capability of NASA.

Finding 2.2: Every NASA atmospheric entry mission has relied on arc jet testing for TPS development.

Finding 2.3: Arc jet facilities provide the only ground-based means of simulating entry heating rates in a reacting flow environment for flight-relevant durations.

Finding 2.4: Ablator design policy currently applies large margins to cover uncertainty in applied loads and material response. Improvements in arc jet diagnostics and test article instrumentation can reduce uncertainties and hence reduce margins

Conclusion 2.1: Existing ablator design policy is incomplete, because some failure modes are not fully understood, so their contribution to system reliability is not quantified. More rigorous examination of material failure modes will enable more defensible mission assurance assessment. Characterization of failure modes is better managed as a material development activity rather than being conducted for individual missions.

Conclusion 2.2: Currently available diagnostic capabilities and instrumentation techniques should be infused into NASA arc jet standard practice. This infusion is best managed as an Agency strategic investment rather than as a programmatic responsibility.

Finding 3.1: Arc jet testing will be needed even for LEO return missions, for which materials already exist and operational experience is in hand. Existing test capability is adequate for this mission type. Capacity requirements to support LEO return are dependent on certification philosophy.

Finding 3.2: NASA and DOD share a mutual reliance on arc jet test capability. Air Force strategic planning relies on the availability of NASA arc jets for testing of hypersonic cruise vehicles. AEDC supports high-shear test conditions that are relevant to NASA.

Finding 3.3: Greater arc jet capability will be required for missions that NASA intends to fly within 30 years. Enthalpy, combined convective-radiative heating, test gas, shear, pressure, turbulence, and model size are among facility features to be addressed.

Finding 3.4: Arc jet facilities that provide more precise control and knowledge of the test conditions, and that have enhanced diagnostics to analyze gas species will benefit the maturation of low-TRL thermal protection concepts and materials technologies.

Finding 3.5: The available TPS choices for flight consideration are not optimal for many missions, and the TPS choices could be increased with improved accessibility and lower cost to use arc jets to develop low TRL concepts.

Finding 4.1: The simulation regimes of the present facilities are sufficient for simulating low Earth orbit entries and certain regimes for Lunar return. However, these facilities are inadequate to fully evaluate the TPS for a Mars or deep space Earth return or certain planet entries such as the Gas Giants.

Finding 4.2: Existing NASA facilities have capability in terms of model size and performance that is not duplicated by DOD or commercial facilities.

Finding 4.3: An inspection of the performance envelopes shows that the JSC capability is basically a subset of the ARC combined capability.

Finding 4.4: Incremental improvements in arc heaters will not deliver the higher test capability needed for more ambitious NASA exploration missions to Gas Giants and human return from Mars.

Finding 4.5: The AEDC power supply cannot operate at low voltage and high current, which is required to power a NASA-type arc heater which operates at low pressure. Therefore, if AEDC is considered as a potential site for operating NASA-type arc heaters, a new power supply will need to be procured.

Conclusion 4.1: NASA must bear the responsibility of designing and developing facilities with improved capability in order to successfully support future NASA missions.

Conclusion 4.2: A focused technology development effort is required to establish the arc jet testing capability required by NASA for future missions.

Finding 5.1: Facilities need to retain operational status, as mothballing an arc jet complex beyond 3 years is tantamount to closure because of the degraded nature of the supporting infrastructure.

Finding 5.2: There are opportunities for improving capacity at both complexes.

Finding 5.3: Staffing levels at both NASA complexes appear to be high.

Conclusion 5.1: Experienced personnel with specialized arc jet training and skills are valuable and irreplaceable. Arc jet facilities must maintain a skilled workforce for safe and reliable operation.

Conclusion 5.2: A more thoughtful approach to building new arc jet capability should consider the larger, long-term Agency mission requirements, as presented in Section 3.0 of this report; integrate the full institutional cost, capability, sustainment, and life cycle into Agency strategic planning; and consider the difference between what is known about arc jet technology and the knowledge needed to advance the state of the art to design an arc jet facility for future capability.

Conclusion 5.3: In an effort to reduce costs, an independent assessment of roles and responsibilities at both the Ames and Johnson complexes should be completed and staffing levels reduced by up to 30%.

Finding 6.1: The return on investment of rehabilitating and upgrading existing complexes, or building a new complex, would be positive. Although arc jet test cost is significant, it is a necessary cost for achieving many of NASA's proposed missions. The total cost of construction is on the same order as one or two subscale flight tests.

Finding 6.2: Arc jet users are cost sensitive, especially commercial customers. They will do more testing if the cost per test is reduced.

Finding 6.3: NASA must have skilled and knowledgeable personnel to develop and operate the next generation of arc heaters.

Finding 6.4: NASA funding and governance of critical institutional technical capability is inadequate. Arc jets have had intermittent funding over several decades, and preventative maintenance has suffered. Strategic Capabilities Assets Program (SCAP) funding to Ames

Research Center has allowed them to work down a backlog of deferred maintenance, resulting in improved availability. JSC has benefited from Shuttle and Constellation program funding, which is expected to go away in 2010.

Conclusion 6.1: NASA needs to make a serious investment in arc jet infrastructure. Investment in the range of $200-500 million in addition to continuing operating costs will be needed in the next 10 years. Resource requirements will be diligently challenged and will limit future technical options.

Conclusion 6.2: Decisions on transition to a new complex need to consider retention and sustainment of highly-qualified staff. Recent cooperation between ARC and JSC has had positive benefits for the Agency and that relationship should be fostered and supported during the transition.

Conclusion 6.3: Cooperative management of JSC and ARC complexes is recommended while a new complex is being developed.

Appendix A: Acronyms

AEDC	Arnold Engineering Development Center
AFRL	Air Force Research Laboratory
AHF	Aerodynamic Heating Facility
AHSTF	Arc Heated Scramjet Test Facility
AJEWG	Arc Jet Evaluation Working Group
AoA	Analysis of Alternatives
ARC	Ames Research Center
ARMD	Aeronautics Research Mission Directorate
ARMSEF	Atmospheric Re-entry Materials and Structures Evaluation Facility
ASAP	Aerospace Safety Advisory Panel
ATP	Aeronautics Test Program
CAD	Computer Aided Design
CAIB	Columbia Accident Investigation Board
CCI	Construction Cost Index
CDI	Critical Design Review
CEV	Crew Exploration Vehicle
CFD	Computational Fluid Dynamics
COTS	Commercial Orbital Transportation Services
CMA	Cryogenic Moisture Apparatus
DARPA	Defense Advanced Research Projects Agency
DOD	Department of Defense
DPLR	Date-Parallel Line Relaxation
EDL	Entry, Descent, and Landing
EDL-SA	Entry, Descent, and Landing – Systems Analysis
ESMD	Exploration Systems Mission Directorate
FIAT	Fan/Inlet Acoustic Technology
HEAT-H1	High Enthalpy Ablation Test Unit H1
HEAT-H2	High Enthalpy Ablation Test Unit H2
HEAT-H3	High Enthalpy Ablation Test Unit H3
HIAD	Hypersonic Inflatable Atmospheric Decelerator
HPITT	Hypersonic Propulsion Integrated Testing Team

HRR	Human Rating Requirements
HyMETS	Hypersonic Material Environmental Test System
ICBM	Intercontinental Ballistic Missile
IHF	Interaction Heating Facility
ISS	International Space Station
ITAR	International Traffic in Arms Regulation
JSC	Johnson Space Center
JTOH	Joint Technology Office on Hypersonics
LaRC	Langley Research Center
LAURA	can't find on NASA Web site
LCAT	(Boeing) Large Core Arc Tunnel
L/D	Lift to Drag ratio
LEAF	Lunar Environment Arc Jet Facility
LEO	Low Earth Orbit
LDR	Lunar Direct Return
LIF	Laser-Induced Fluorescence
MDR	Mars Direct Return
MHD	Magneto Hydrodynamics
MOLA	Mars Orbiter Laser Altimeter
MSL	Mars Science Laboratory
NASA	National Aeronautics and Space Agency
NOx	Nitrogen Oxide
NPAT	National Partnership for Aeronautical Testing
OCE	Office of the Chief Engineer
OD	Occupancy Days
PDR	Preliminary Design Review
PICA	Phenolic Impregnated Carbon Ablator
PTF	Panel Test Facility
RHTF	Radiant Heat Transfer Facility
RCC	Reinforced Carbon Carbon
RTF	Return to Flight
SLA	Super Lightweight Ablator
SMD	Science Mission Directorate

SOMD	Space Operations Mission Directorate
STAB	Structural and Technical Analysis Board
STS	Space Transportation System (Program)
SVS	Steam Vacuum System
TFD	Turbulent Flow Duct
TRL	Technical Readiness Level
TPS	Thermal Protection System
TPTF	Truncated Panel Test Facility

Appendix B: Glossary

Availability: the planned uptime of a facility or asset; the definition of availability is generally the sum of planned uptime (operational schedule and planned downtime) divided by the sum of planned uptime plus unplanned downtime.

Capability: the set of test conditions that can be achieved at an individual facility or an entire complex or complexes.

Capacity: full throughput utilization of a facility or complex, a property of a complex and associated staff.

Complex: a location with one or more facilities that includes the supportive infrastructure (power, boilers) plus the arc jet(s).

Facilities Maintenance: The recurring day-to-day work required to preserve real properties (land, buildings, structures, utility systems, collateral equipment, and other permanent improvements) in such a condition that they may be used for their designated purpose over an intended service life. It includes the cost of labor, materials, and parts. Maintenance minimizes or corrects wear and tear, forestalling major repairs.

Facility: an arc jet tunnel.

Greenfield: new construction of a complex.

Maintenance: the recurring actions, funded through an annual and planned budget, necessary to keep equipment and systems operational. Includes routine care, preventive maintenance (or predictive maintenance), trouble calls, and repair. [NPR 8831.1E]

Operations: the recurring actions, funded through an annual and planned budget, necessary to exercise equipment and systems for some pre-determined purpose. For the Arc Jets, "operations" involves simultaneously running the equipment and systems required to create the thermal simulation environment necessary to capture data on the performance of a calibration probed or test model under those test conditions.

Recapitalization: the renewal and modernization of infrastructure to ensure its ability to meet current and future program and institutional requirements. Recapitalization projects typically are large (costs and schedule) and are designed to last 20-30 years.

Revitalization: the renewal and modernization of research equipment and systems to ensure its ability to meet current and future program requirements. Revitalization projects typically are small to medium (costs and schedule) and are designed to last 3-10 years.

Appendix C: References

- ARC, Basis of Need document for ARC boiler, FPMD Project #15959, July 30, 2009
- Ares Corporation, "Arc-Heated Test Facility Investment and Risk Reduction Analysis of Alternatives Study for Orion Heat Shield Development", Report No. 063010514-001, May 2007.
- Austin, Howard, "Summary of Current NASA Facility Maps," Eloret Corp., NASA Ames Research Center.
- Bowman, Keith, Air Force Research Laboratory: Arc Jet Testing Perspective, HQ presentation, February 16, 2010
- Curry, Don, Past Experience Using Arc Jet Testing for Human Spacecraft TPS. JSC presentation, February 8, 2010
- Del Papa, Steven, "Atmospheric Reentry Materials and Structures Evaluation Facility," presented February 8, 2010.
- Kardell, Matt, "Boeing LCAT Facility Arc-Jet Test Capability", Presentation to AJEWG, February 16, 2010.
- Laub, Bernie, et al., Past Experience: A Different Perspective, ARC presentation, February 10, 2010
- McDonald, Mark, Lunar re-Entry eXperiment (LEX) Study Team Final Report, March 2008
- Munk, Michelle, User Briefing: Future Missions, JSC presentation, February 8, 2010
- NASA, "Entry Thermal Protection," NASA SP-8014, August 1968
- Prabhu, Dinesh, Modeling as an Alternative or Augmentation to Testing, ARC presentation, February 17, 2010
- Raiche, George, "NASA Ames Arc Jet Complex Overview", Presentation to AJEWG, February 10, 2010.
- Shepard, Charles E. and Wm. C.A. Carlson, "Upgrading of NASA-Ames High-Energy Hypersonic Facilities – a Study," Contract No. NCC2-503, February 1988.
- Smith, Mark D., "AEDC Arc Jet Facility Capabilities and Test Environments", Presentation to AJEWG, February 16, 2010.
- Steltzner, Adam, Mars Exploration Program Arc Jet Use Profile, ARC presentation, February 10, 2010
- Venkatapathy, Ethiraj, "Arc Jet Capabilities Needed for Future Venus and Outer Planet Missions", Presentation to AJEWG, February 10, 2010.
- Wright, Michael, "Anticipated Arc Jet Usage, Thermal Protection System Technology Development in Support of High Mass Planetary Entry and Safe Return of Crew to Earth," Presentation to AJEWG, February 10, 2010.

Appendix D: Study Information

A.1 Arc Jet Evaluation Working Group

- Mike Ryschkewitsch, NASA HQ, Chief Engineer, Study Lead
- Anthony Calomino, NASA GRC, Working Group Lead
- Mike Mastaler, NASA ATP, Study Executive
- Hal Bell, NASA HQ, Planning and Execution Support
- Walter Bruce, NASA LaRC, Arc Jet Subject Matter Expert
- Peter Gage, Neerim Corp, Arc Jet Subject Matter Expert
- Dennis Horn, AEDC (retired), Arc Jet Subject Matter Expert
- Don Rigali, Sandia National Laboratory (retired), Arc Jet Subject Matter Expert
- Judith Robey, NASA HQ, Independent Program and Cost Evaluation
- Jerry Wahlberg, NASA LaRC and NC State University (retired), Arc Jet Subject Matter Expert
- Calvin Williams, NASA HQ, Facilities Subject Matter Expert
- Linda Voss, Dell-Perot, Report Editor
- Glen Campbell, Dell-Perot, Planning and Execution Support
- Judith Jacobs, Dell-Perot, Support

A.2 Review Schedule and Presentations

Work Statement	November 2009
First Draft Report Due	February 26, 2010
Final Report Due	March 26, 2010

Meetings

Team On-site at JSC	February 8 and 9, 2010
Team On-site at ARC	February 10 and 11, 2010
Team On-site at HQ	February 16 and 17, 2010
Team On-site at HQ	March 18 and 19, 2010

Presentations

Team On-site at JSC   February 8 and 9, 2010

- Stan Bouslog, Why Do TPS Engineers Need Arc Jets?
- Andrew Chambers and John Crumpler, SpaceX Thermal Protection Systems
- Don Curry, Past Experience Using Arc Jet Testing for Human Spacecraft TPS
- Steven Del Papa, Atmospheric Reentry Materials and Structures Evaluation Facility
- Scott Durrant, Arc Jet for TPS Design
- James Reuther, presenter, Orion, David Hash, Arc Jet Facility Capabilities and Lunar Return Certification Requirements
- E. Heddles, Orbiter TPS Development and Certification Testing At the NASA/JSC 10 MW Atmospheric
- E. Heddles, Alternate Use of the Atmospheric Re-entry Materials and Structures Evaluation Facility (ARMSEF) Due to NASA Program Changes
- John Kowal, Space Shuttle Orbiter and Orion TPS Arc Jet Testing
- John Kowal, Orion TPS Advanced Development Project (ADP) Arc Jet Testing
- Chris Madden, JSC Arc Jet - Introduction
- Chris Madden, EHTF Background
- Jim Milhoun, JSC Arc Jet Tour, Summary of Informal Presentation on Historical Background of JSC Arc Jets, February 2010
- Michelle Munk, User Briefing: Future Missions
- Joe Riccio, JSC Experimental Heat Transfer Facilities and Upgrade Status
- W.C. Rochelle, Orbiter TPS Development and Certification Testing at the NASA/JSC 10 MW
- Jeremy Vander Kam and E. Heddles, Facility-to-Facility PICA Recession Comparisons (TP2 vs. IHF/AHF)
- Jeremy Vander Kam, Orion DAC 3 ISS Ascending Node Trajectories

Evaluation of Arc Jet Facilities Report

- Reentry Materials and Structures Evaluation Facility

Team On-site at ARC   February 10 and 11, 2010

- Jim Arnold, Past Experience with ARC Arc Jet Complex and Benefits of their Being Integrated with TPS R&D and Aerothermodynamics
- Neil Cheatwood, Development of Flexible TPS for Hypersonic Inflatable Aerodynamic Decelerators
- Bernie Laub, Past Experience: A Different Perspective
- Dinesh Prabhu, Modeling as an Alternative or Augmentation to Testing
- George Raiche, Ames Arc Jet Complex Overview Power Supply Configuration Update
- George Raiche, NASA Ames Arc Jet Complex Overview
- Adam Steltzner, Mars Exploration Program Arc Jet Use Profile
- Jeremy Vander Kam, Orion TPS I/O, Reliability as a Function of Test Quantity
- Raj Venkatapathy, Arc Jet Capabilities Needed for Future Venus and Outer Planet Missions
- Mike Wright, Anticipated Arc Jet Usage: Thermal Protection System Technology Development in Support of High Mass Planetary Entry and Safe Return of Crew to Earth (ARMD/ETDP Technology Programs)

Team On-site at HQ   February 16 and 17, 2010

- Keith Bowman, USAF, Air Force Research Laboratory (AFRL): Arc Jet Testing Perspective
- Matt Kardell, Boeing LCAT Facility Arc Jet Test Capabilities
- Mark McDonald, LEX Study Report
- George Raiche, Ames Arc Jet Complex Overview: Additional Requested Information
- Steve Rickman, Considerations for the Arc Jet Evaluation Working Group (AJEWG)
- Mark Smith, AEDC Arc Jet Facilities Overview

Team On-site at HQ   March 18 and 19, 2010
- Christopher Madden and George Raiche, Cooperative Arc-Jet Management Proposal Brief to NASA AJEWG

A.3 Other Reference Documents

- ARC, Basis of Need document, July 30, 2009
- ARC Basis of Need Document for Boiler, January 29, 2009
- ARC Boiler Replacement Facility Project Document, September 2, 2009
- ARC/JSC Facility Assessment Study for Aeronautics Test Program and Shared Capability Asset Program-Final Report, January 2009
- ARC Master Plan, April 2009
- ARC: The Arc Jet Complex (technical brochure), 2000
- Don Ellerby, Orion TPS Thermal Test Plant Facility Status, October 26, 2009
- Steve Frankel, ARC Boiler Replacement Project Paper, April 7, 2006
- Joe Hartman, ARC Complex Overview, March 2, 2006
- Impacts to Agency Mission of Closure of JSC Arc Jet Facility
- JSC Atmospheric Re-entry Materials, October 12, 2009
- JSC Boiler System Upgrade – Economic Analysis, April 24, 2009
- JSC Facility Project Document Describing Boiler, August 31, 2009
- JSC Mechanical and Boiler Systems Upgrade – Economic Analysis, April 4, 2009
- JSC Mechanical and Boiler Systems Upgrade - Economic Analysis, April 24, 2009
- Jacobs Facility Assessment Study for Aeronautics Test Program and SCAP: Final Report, January 2009
- Chris Madden and Steve Rickman, JSC Infrastructure Improvement Plan, September 8, 2008
- Mark McDonald, Lunar re-Entry eXperiment (LEX) Study Team Final Report, March 2008
- Optimization of Agency's Arc Jet Capacity, Utilization, and Cost
- George Raiche, ARC Complex Capability Description, September 2009
- George Raiche, ARC Infrastructure Planning and Status, September 1, 2009
- George Raiche and Peter Chan, Arc Jet Capability Revitalization Options, January 2010
- SCAP Issues Paper for FY2011-2016

- D. Mark Smith, AEDC Arc Jet Facility Capabilities and Test Environments, February 16, 2010
- Strategic Capabilities Assets Program Issue Paper
- Strategic Capabilities Assets Program: An Overview, November 30, 2009
- Sverdrup Arc-Heated Facility Investment and Risk Reduction Analysis of Alternatives Study for Orion Heat Shield Development

Appendix E: AJEWG Team Bios

Arc Jet Evaluation Working Group

- Michael (Mike) Ryschkewitsch, Lead
  Chief Engineer, Office of the Chief Engineer
  NASA HQ
  (202) 358-1823

As Chief Engineer, Mr. Michael Ryschkewitsch is responsible for the overall review and technical readiness of all NASA programs. The Office of the Chief Engineer ensures that the agency's development efforts and mission operations are planned and conducted on a sound engineering basis with proper controls and management of technical risks.

Previously, Mr. Ryschkewitsch served as the Deputy Director for NASA's Goddard Space Flight Center in Greenbelt, Md., and Director of the Applied Engineering and Technology Directorate at Goddard. He joined the Center in 1982 as a cryogenics engineer to work on the Cosmic Background Explorer mission. Between those jobs, Ryschkewitsch held several management positions and supported projects from the first servicing mission of the Hubble Space Telescope in 1993 to the Aeronomy of Ice in the Mesosphere mission launched in April 2007.

Mr. Ryschkewitsch earned his B.S. in Physics from the University of Florida in 1973 and his Ph.D. from Duke University in 1978.

Arc Jet Evaluation Team

- Dr. Peter Gage
  Neerim Corporation (Mountain View, CA)
  (650) 646-2897 (office)
  (650) 269-9328 (cell)
  pgage@neerimcorp.com

Dr. Peter Gage is the Founder and President of Neerim Corp., Mountain View, California, which provides systems engineering and design support to government and industry.

Prior to that, Dr. Gage was a Senior Engineer at Valador Inc., where he was the Thermostructural Analysis Team Lead for NASA's Advanced Development Project for Orion heat shield. He also supported NASA's Simulation-Assisted Risk Assessment (SARA) project on Ares 1 Abort Effectiveness. Previously, Dr Gage worked on-site at NASA Ames, where he was Technical Director of ELORET Corp.

Dr. Gage has a proven track record in vehicle design for NASA, primarily on Entry, Descent, and Landing, and with recent experience on launch vehicles and mission architecture. He also has long involvement in development of software and systems to support aerospace design teams, having worked on NASA's Advanced Engineering Environment and several similar programs.

Dr. Gage is an Associate Editor for Journal of Spacecraft and Rockets. He is an Associate Fellow of AIAA, a member of AUVSI and a member of INCOSE. He earned a Ph.D. in Aeronautics and Astronautics from Stanford University in 1994.

- Dennis Horn
  AEDC (retired; Tennessee)
  (931) 461-0262 (home)
  931-581-9901 (cell)
  horndd@lighttube.net, horndd2@lighttube.net

Dennis Horn is retired (1997) from the Air Force Arnold Engineering Development Center (AEDC) and provides consulting work for NASA, AEDC, and Lytec, LLC.

Dennis is an engineering specialist with over 50 years experience in the field of hypersonics and aerothermodynamics. During his 38 year career at the AEDC, he worked 35 years developing electric arc heaters used to simulate hypersonic and re-entry flight conditions. He was the principal investigator for all segmented arc heaters developed and operated at the AEDC. These arc heaters operate at pressure levels up to five times higher than other segmented arc heaters worldwide. The largest heater has operated at power levels up to 68 MWs. His experience at the AEDC includes arc heater performance predictions, hardware development, including segment design and fabrication, and arc stability studies. Dennis was named an AEDC Fellow in 1989 by the U. S. Government for his contributions in advancing the "state of the art" in segmented arc heater technology.

Dennis earned a B.S. in Mechanical Engineering from the University of Illinois in 1957 and an M.S. in Theoretical and Applied Mechanics from the University of Illinois in 1958.

- Don Rigali
  Sandia National Laboratory (retired; New Mexico)
  (505) 856-1620 (home)
  (505) 270-5191 (cell)
  djrigali@att.net

Mr. Donald (Don) Rigali is retired (1998) from the Department of Energy's Sandia National Laboratories (SNL) and is a consultant to SNL, the Air Force Research Lab (AFRL), Ball Aerospace and Tech. Corp., and NASA, including the Columbia Accident Investigation Board (CAIB).

Before retiring, Don was the Director for the Aerospace Systems Development Center at SNL where he directed work in the areas of National Missile Defense, Theater Missile Defense, smart targeting, hypersonic weapons and technology, and access to space. Under Don's direction, the Center became a recognized, national leader for conducting rapid, proof-of-principle/pre-prototype demonstrations of re-entry, weapons, countermeasures, space, rocket, aerodynamic flight, and handicap mobility systems. Don provided program direction and oversight of reimbursable programs for various agencies, including the Ballistic Missile

Defense Organization, U.S. Army Space & Strategic Defense Command, Advanced Research Projects Agency, U.S. Navy Strategic Systems Project Office, Defense Nuclear Agency, and U.S. Air Force Phillips Lab. Other work areas in Don's history include re-entry systems research and development; high speed vehicle research and development; novel reentry systems penetration aides; SWERVE (hypersonic glide vehicle flight tested to M=13); high-altitude sounding rocket research and development; high speed (M=7) rocket sled design, development, and testing; and wind tunnel testing.

Don earned a B.A. in Liberal Arts from the University of St. Thomas in 1955, a B.S. in Aeronautical Engineering from the University of Notre Dame in 1957, and an M.S. in Mechanical Engineering from the University of New Mexico in 1968.

- Dr. Gerald (Jerry) Walberg
  NASA Langley Research Center (retired)
  (757) 723-4607
  gwalberg@gmail.com

Dr. Gerald (Jerry) Walberg is retired (2009) from Walberg Aerospace, a research company specializing in entry aerothermodynamics, trajectory optimization, and planetary mission analysis. After establishing Walberg Aerospace in 1999, Dr. Walberg consulted for the NASA Langley Research Center on the Revolutionary Aerospace Concepts Program and carried out re-entry safety analyses on the Stirling Radioisotope Power System for Teledyne Energy Systems and the Multimission Radioisotope Thermoelectric Generator System for Boeing/Rocketdyne.

From 1957 to 1989, Dr. Walberg was employed at Langley Research Center where he held positions ranging from Research Engineer to Deputy Director for Space. Following retirement from NASA, he taught at the NASA/George Washington University Joint Institute for Advancement of Flight Sciences and then at North Carolina State University, where he was the Director of the Mars Mission Research Center in the Department of Mechanical and Aerospace Engineering. In this capacity, he provided overall administration for the Center programs, led the Center's mission analysis and design activities, and taught graduate and undergraduate courses in hypersonic aerodynamics, celestial and orbital mechanics, and spacecraft design.

As Deputy Director for Space for Langley Research Center, Dr. Walberg planned, directed, and coordinated the space-related research, science, and technology activities of the Center. As Chief of the Space Systems Division there, he planned and directed the Division's support of current and planned NASA missions and emphasized the aerothermodynamics of advanced space vehicles, space energy systems, the development of computer-aided analysis techniques, and systems analyses of spacecraft for future missions. In 1979, Dr. Walberg was selected to participate in the NASA Executive Development Program at NASA Headquarters. In this role, he was responsible for overall long-range planning for space transportation technology, and he served as Deputy Chairman for the F.Y. 1981 OAST Space Technology Assessment. As Head of the Thermodynamics and Combustion Section, the Gas Radiation Section, and the Aerothermodynamics Branch at Langley, Dr. Walberg was responsible for directing a broad-based program of theoretical and experimental research on high-temperature gas dynamics and ablation. He played a lead role in the analysis and testing of the Apollo heat shield. He led a team that developed some of the first rigorous analyses of radiatively-coupled flow fields, and he supported the Viking, Pioneer Venus, and Galileo Probe missions.

From 1970 to 1987, Dr. Walberg served as Chairman of the Re-entry Sub-Panel of the Interagency Nuclear Safety Review Panel. In this capacity, he played a key role in the nuclear safety reviews for all U.S. missions that involved the use of nuclear power sources in Earth orbit and beyond, directed the reviews of all re-entry analyses, and participated in the development of the Safety Evaluation Reports for the Pioneer, Viking, LES, Voyager, and Galileo missions.

Arc Jet Evaluation Technical Experts

- Walter (Walt) Bruce
Langley Research Center; NASA Engineering and Safety Center

(757) 864-7024 (office)
(757) 871-0377 (cell)
walter.e.bruce@nasa.gov

Mr. Walter (Walt) Bruce is a Senior Thermal Engineer in the Langley Research Center's Systems Engineering Directorate. Walt has extensive experience with hypersonic thermal testing including test planning for the Flexible Thermal Protection System Project in the Ames Panel Test Facility. He was the arc jet test lead for the Mars Science Laboratory Entry, Descent and Landing Instrumentation (MEDLI) Project, where he tested 156 models in the Boeing LCAT Facility; and the technical lead for the Falcon leading edge material arc jet test in the AEDC H2 Facility. He served on the MSL Heat Shield Tiger Team and was the arc jet test lead for the Hyper-X (X-43A) nose leading edge, where he tested two models in the AEDC H2 Facility. He was also the arc jet test lead for the Shuttle leading edge patch repair, where he tested 85 models in the Boeing LCAT Facility and 22 models in the ARC IHF.

Prior to this, Walt worked at AEDC in the arc jet facilities, where he performed arc heater technology and development testing and managed the development of the AEDC H3 arc heater facility, a large, high-pressure, segmented arc heater. He also performed technology testing in the Aerospatiale multi-arc heater JP-200 facility and the High Pressure (HP) arc facility in Bordeaux, France.

Walt received his B.S. in Mechanical Engineering from North Carolina State University in 1983 and his M.S. in Mechanical Engineering from North Carolina State University in 1985.

- Dr. Anthony Calomino, Lead
Glenn Research Center
(216) 433-3311 (office)
(216) 513-0489 (cell)
anthony.m.calomino@nasa.gov

Dr. Anthony Calomino spent 20 years as a research engineer within the Mechanics and Life Prediction Branch at the Glenn Research Center. Dr. Calomino's primary research interests and experience have been in damage analysis and materials behavior modeling for high temperature aerospace materials, including refractory materials and composites used in commercial, military, and space applications. He worked with the Columbia Accident Investigation Board and Return to Flight investigating the mechanical behavior of the composite leading edge. For the last four years, Dr. Calomino has served as Hypersonics Materials and Structures Lead Engineer within the Fundamental Aeronautics Program's Hypersonics Project under the Aeronautics Research Mission Directorate.

Dr. Calomino received his B.S. in Structural Engineering from the University of Colorado at Boulder, completed an M.S. in Engineering Mechanics at Case Western Reserve University in Cleveland, and was awarded a Ph.D. in Materials Science from Northwestern University in Evanston.

- Michael (Mike) D. Mastaler, Working Group Technical Representative
  ARMD Aeronautics Test Program; NASA HQ
  (202) 358-1105 (office)
  (202) 236-0437 (cell)
  michael.d.mastaler@nasa.gov

For the last 20 months, Michael (Mike) Mastaler has worked as the HQ Liaison for the Aeronautics Research Mission Directorate's Aeronautics Test Program (ATP). He is a permanent employee of the Langley Research Center on detail to NASA HQ. In this position, he is responsible for all ATP-related activities at HQ and for facilities-related activities for ARMD.

Prior to that, Mike was the Deputy Director for Business Management, Center Operations Directorate, where he was responsible for the management of $120 million per year budget associated with wind tunnel operations, facility capital investments and sustainment, and the day-to-day operation of the Langley Research Center. Mike has 26 years of experience with NASA and has served as a process and fluid systems engineer, Project Manager, and as the head of process engineering, mechanical engineering and design, and resources management groups.

Mike earned a B.S. in Mechanical Engineering from Old Dominion University in 1983, an M.S. in Mechanical Engineering from Old Dominion University in 1989, and an MBA from the College of William and Mary in 2001.

- Judith (Judee) Robey
  Program Analysis and Evaluation
  NASA HQ
  (202) 358-0823 (office)
  (443) 875-8913 (cell)
  judith.l.robey@nasa.gov

Judith Robey is currently a study manager in the Program Analysis and Evaluation Office and manager of the Agency Mission Planning Manifest. Ms. Robey worked for 16 years on the International Space Station Program (ISS). This included defining requirements for Destiny and International Laboratories for Physical Science research. Additionally, she was the program manager for the development of research instruments for several Spacelab missions and the fluids, combustion and materials science racks currently on the ISS. She also conducted negotiations and agreements with international partners on cooperative ISS research. Prior to this Ms. Robey worked for 6 years as a research scientist at the Jet Propulsion Laboratory. Ms. Robey received her degree in physics at the University of California, Los Angeles.

- Calvin Williams
  Facilities Engineering and Real Property Division
  NASA HQ
  (202) 358-2322 (office)
  (301) 442-2380 (cell)
  calvin.williams@nasa.gov

Calvin Williams is the Chief of the Planning and Real Estate Branch in the Facilities Engineering and Real Property Division, where he manages the Agency's real property portfolio and Agency master planning efforts. He has recently assumed the responsibility to manage the Strategic Capabilities Asset Program Branch in addition to his other responsibilities.

Mr. Williams joined NASA in 2002, where he served as the Code OJX's Lead Program Manager for Code R (Aeronautics) Centers. His responsibilities included managing the Construction of Facilities projects at NASA's Venters. He also served as the Sustainable Design Champion at NASA Headquarters, leading the Agency's efforts in sustainable designs and certifications under the U.S. Green Building Council's Leadership in Energy and Environmental Designs program.

Before coming to NASA, Mr. Williams worked for the National Institutes of Health (NIH) in Bethesda, Maryland for 12 years. He held several positions, including program manager, business manager, and management representative for ISO 9001 in the Design, Construction, and Alterations Branch. Mr. Williams managed both technical and administrative staff in support of design and construction activities at the NIH. As the Design Project Director of the Infrastructure Modernization Program, Mr. Williams managed the Master Utility Plan for upgrading the entire major utility infrastructure for the NIH Bethesda Campus.

Prior to joining NIH, Mr. Williams worked for the Department of the Navy. He worked in facilities engineering positions for both shore activities and on nuclear submarines. He held the position of Program Manager at the Chesapeake Division of the Naval Facilities Engineering Command, responsible for managing energy and other facility engineering projects. As a Program Manager for the Naval Sea System Command, Mr. Williams was responsible for testing, monitoring, and extending the operating life of HVAC systems onboard the Navy's nuclear class submarines.

Mr. Williams graduated from Howard University with a B.S in Mechanical Engineering. He also obtained an M.S. in Engineering from the Johns Hopkins University.

- Hal Bell
  Office of the Chief Engineer; NASA HQ
  (202) 358-1040 (office)
  (202) 391-3935 (cell)
  harold.m.bell@nasa.gov

- Linda Voss
  Report Editor
  Dell Perot Systems
  (703) 524-3554 (phone and fax)
  (703) 867-1817 (secondary number)
  Linda.Voss@psgs.com

Appendix F: Current Arc Jet Capabilities

This appendix assesses and compares the performance capabilities of the overall arc jet complexes support equipment and individual arc jet facility capabilities.

Ames Research Center Arc Jet Complex

The ARC arc jet complex consists of eight available test bays, all with vacuum exhaust capability located in two separate buildings as shown in Figure F-1. Four of the test bays contain operational arc jet facilities: IHF, AHF, PTF, and TFD. All of the test bays share a common steam-ejector driven vacuum system, a water cooling system, high-pressure gas system, and other auxiliary systems. Two power supplies are available—a nominal 60MW power supply and a 20MW power supply. The IHF facility uses the 60-MW power supply while the other three facilities use the 20-MW power supply.

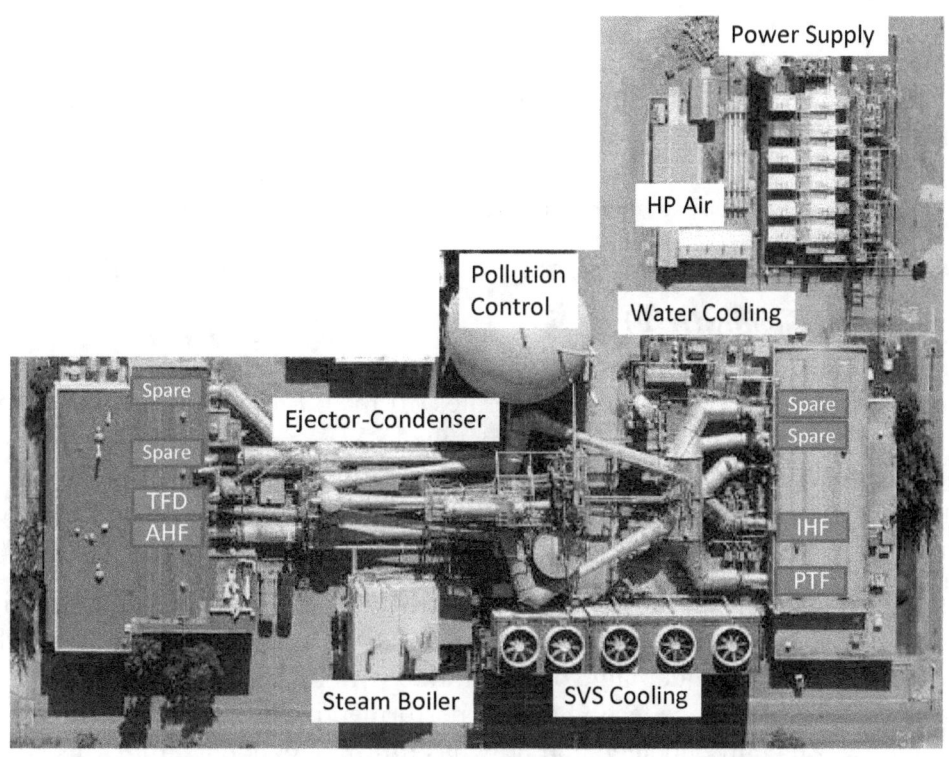

Figure F-1. Aerial View of Ames Arc Jet Complex.

The current configurations for each of the four facilities are presented in Table F-1. Note that the AHF facility can be operated with a segmented or Huels type heater, resulting in different overall test capabilities.

Table F-1. ARC Arc Jet Complex Active Configuration Summary

*Heating Rate is a cold wall, fully catalytic value on a 4-inch diameter hemisphere

Each of the four active arc jet facilities are described in detail in the sections below.

Interaction Heating Facility (IHF)

The IHF uses a segmented type arc heater with multiple ring electrodes on either end of the heater as shown in Figure F-2. The heater operates off the 60-MW power supply; however, the typical maximum operating power is approximately 45 MW. The facility is equipped to run off of air or nitrogen test gas and uses argon shield gas over the electrodes to help with arc attachment on the electrodes. The heater has cold-air injection capability on the downstream end of the heater, prior to the nozzle entrance, to provide a lower-enthalpy test gas.

A 10-degree half-angle conical nozzle with five individual segments is available (Figure F-3), which results in a range of nozzle exit diameters from 6 in. to 41 in. Several throat diameters are available, which results in a wide range of area ratios with minimal hardware. Free-jet stagnation type testing or free-jet shear type testing can be performed using the conical nozzles. The model injection system was recently updated and now has four independent sting arms similar to the ones shown in Figure F-4. A 8-in. by 32-in. semi-elliptic nozzle is also available, which is typically used for large panel (up to 24 in. by 24 in.) shear type testing. Performance envelopes for each of these types of testing are shown in Figure F-5 through Figure F-9.

Figure F-2. Ames Interaction Heating Facility (IHF).

Flow

Figure F-3. 10-degree Conical Nozzle Segments for IHF.

Figure F-4. Model Injection System Showing Two Sting Arms.

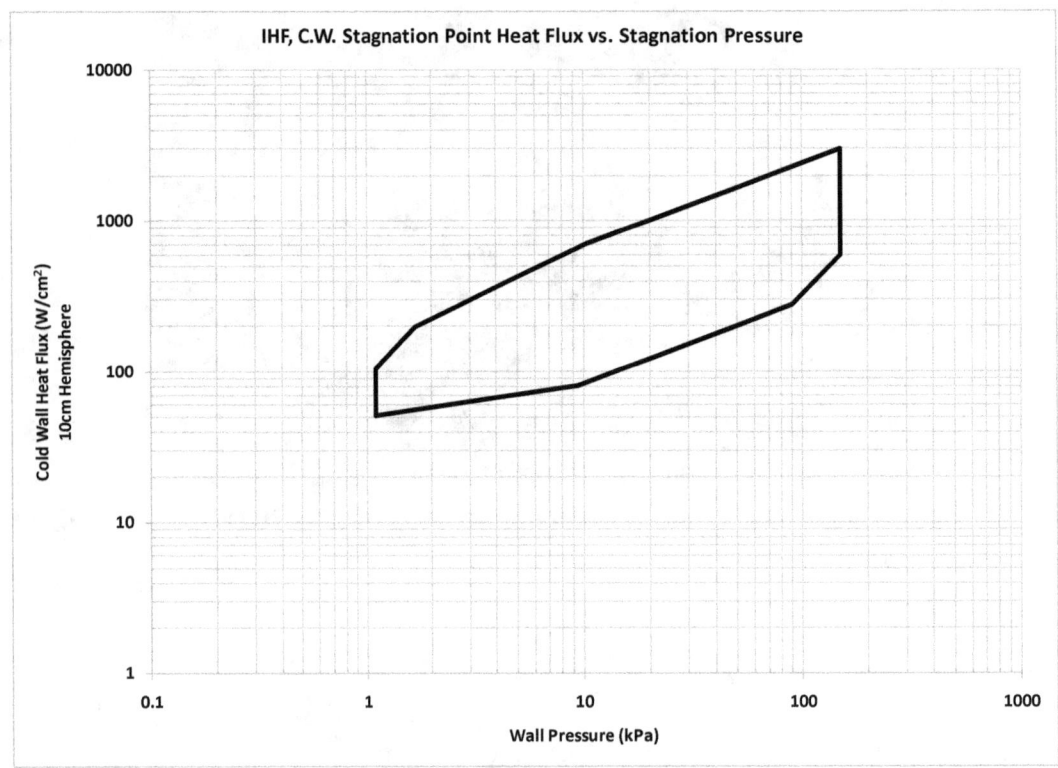

Figure F-5. IHF Stagnation Performance for Cold Wall Heat Flux vs. Pressure.

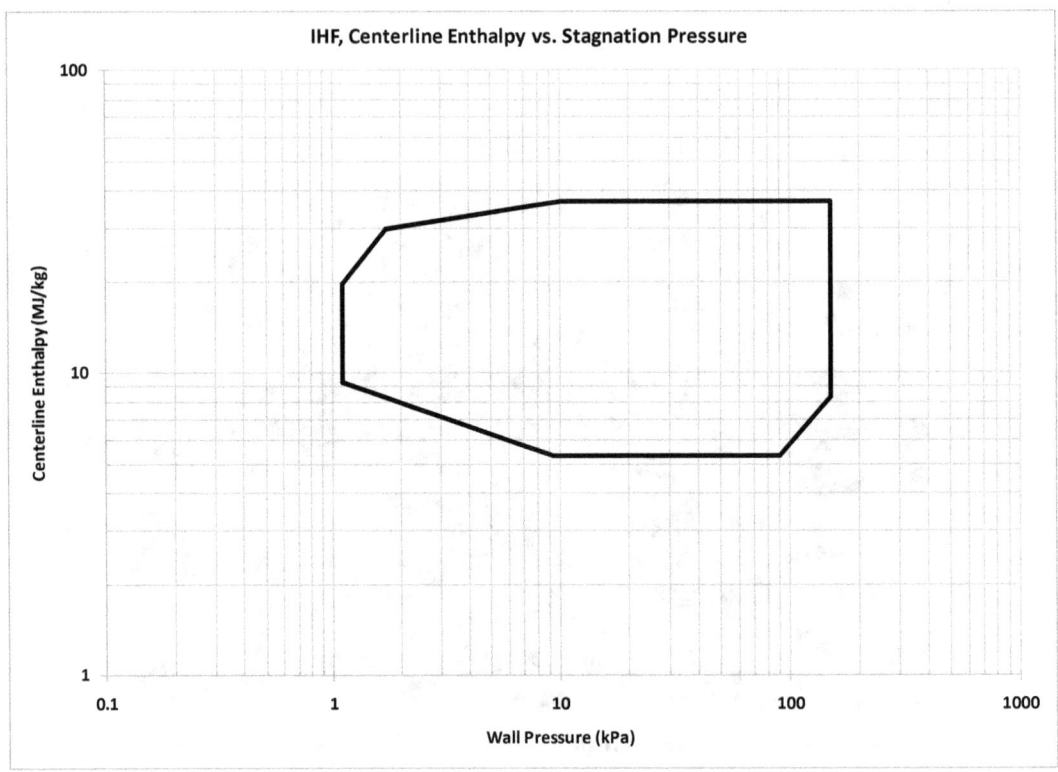

Figure F-6. IHF Stagnation Performance for Centerline Enthalpy vs. Pressure.

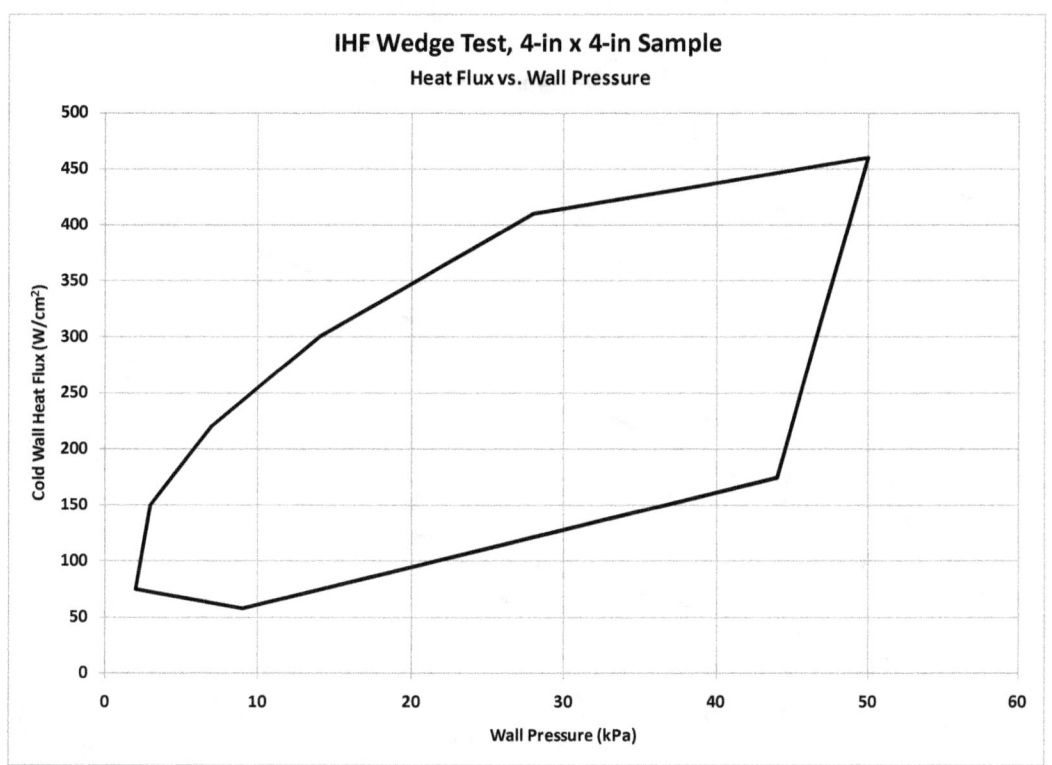

Figure F-7. IHF Wedge Test Capability on a 4-in x 4-in Sample, Heat Flux vs. Wall Pressure.

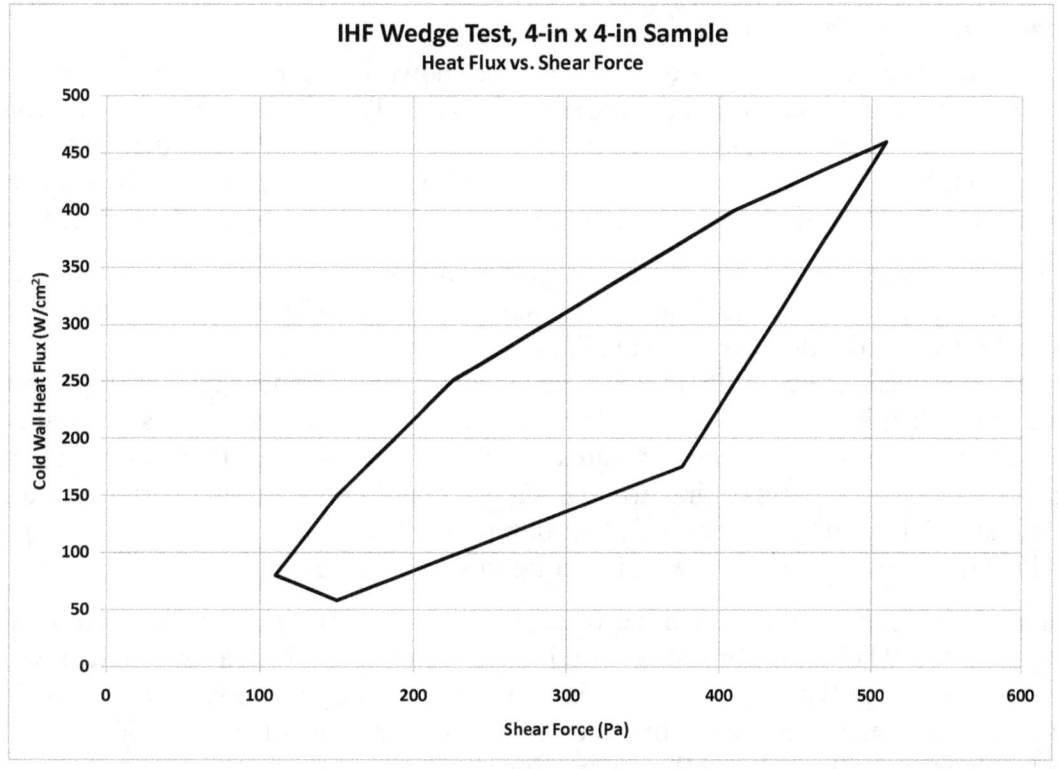

Figure F-8. IHF Wedge test Capability on a 4-in x 4-in Sample, Heat Flux vs. Shear Force.

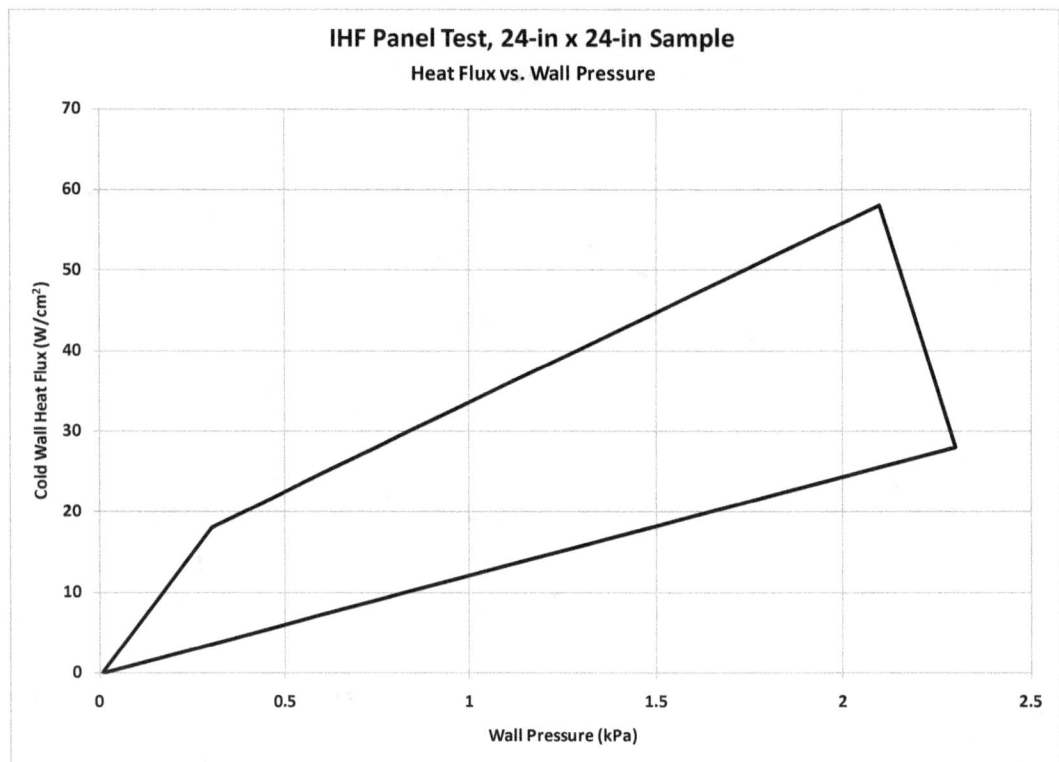

Figure F-9. IHF Panel Test Performance on a 24-in x 24-in Sample, Heat Flux vs. Wall Pressure.

Aerodynamic Heating Facility (AHF)

The AHF can operate with a segmented arc heater, as shown in Figure F-10, or with a Huels arc heater. This results in different test capabilities as shown in Table F-1. The facility is typically operated with the segmented heater, which has multiple ring electrodes on either end of the heater. The heater, segmented or Huels, operates off of the 20-MW power supply and can operate using air or nitrogen. The segmented heater uses argon shield gas over the electrodes.

An 8-degree half-angle conical nozzle with five individual segments, similar to the IHF conical nozzle shown in Figure F-3, is available. This results in a range of nozzle exit diameters from 12 in. to 36 in. Several throat diameters are available, which results in a wide range of area ratios with minimal hardware. Free-jet stagnation type testing or free-jet shear type testing can be performed in this facility. Stagnation models can be as large as 8-in. diameter and wedge models as large as 26 in. by 26 in. Five sting arms are available on a linear injection system, as shown in Figure F-11, with one swept type sting arm, which is typically used for an instrumentation probe. The five-arm model injection system is fully programmable and has the capability of injecting the model nearly vertical into the flow or injecting in a transverse motion.

The segmented heater operates at chamber pressures from 1 to 9 atm and bulk enthalpy levels from 500 to 14,000 BTU/lbm. Convective heating rates can range from a low value of 0.05 BTU/ft^2-sec with wedge type testing to a high value of 225 BTU/ft^2-sec with the 12-in. diameter conical nozzle. The Huels type heater operates at chamber pressures from 1 to 40 atm and bulk enthalpy levels from 1,500 to 4,500 BTU/lbm. Convective heating rates are approximately the same as the segmented because the surface pressures on the model tend to be higher, which offsets the lower enthalpy typically associated with a Huels type heater.

Figure F-10. Ames Aerodynamic Heating Facility (AHF)

Figure F-11. AHF model Injection System.

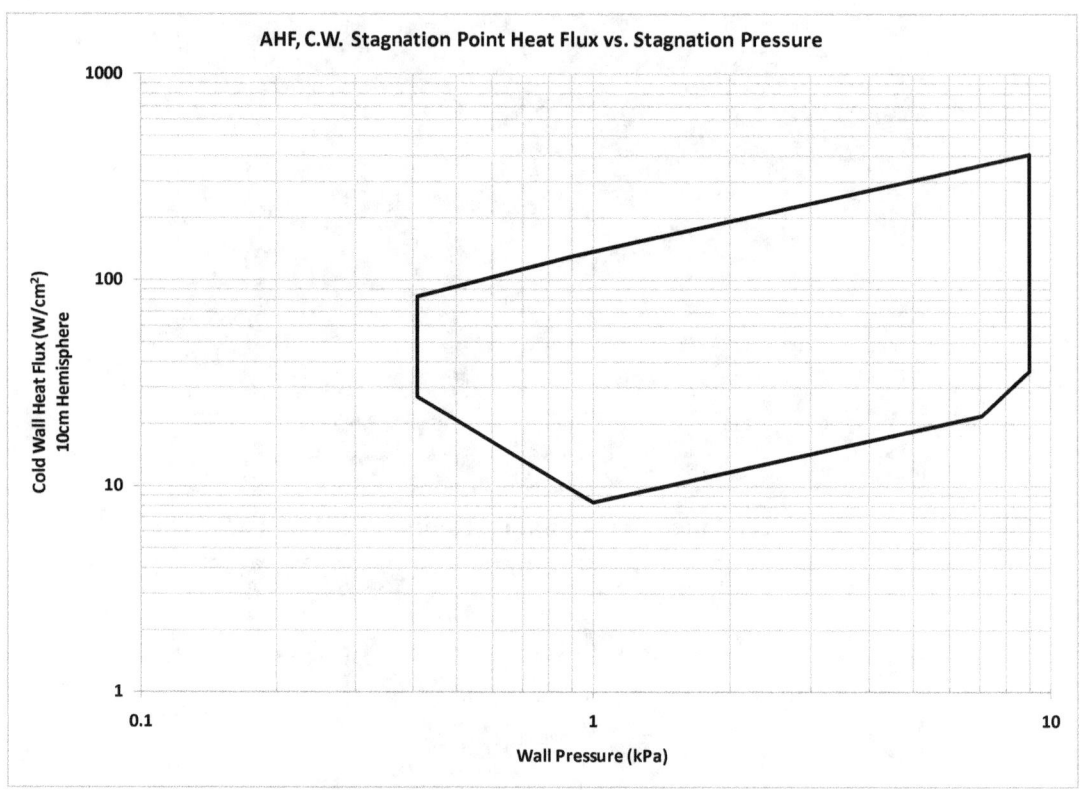

Figure F-12. AHF Stagnation Performance, Heat Flux vs. Pressure.

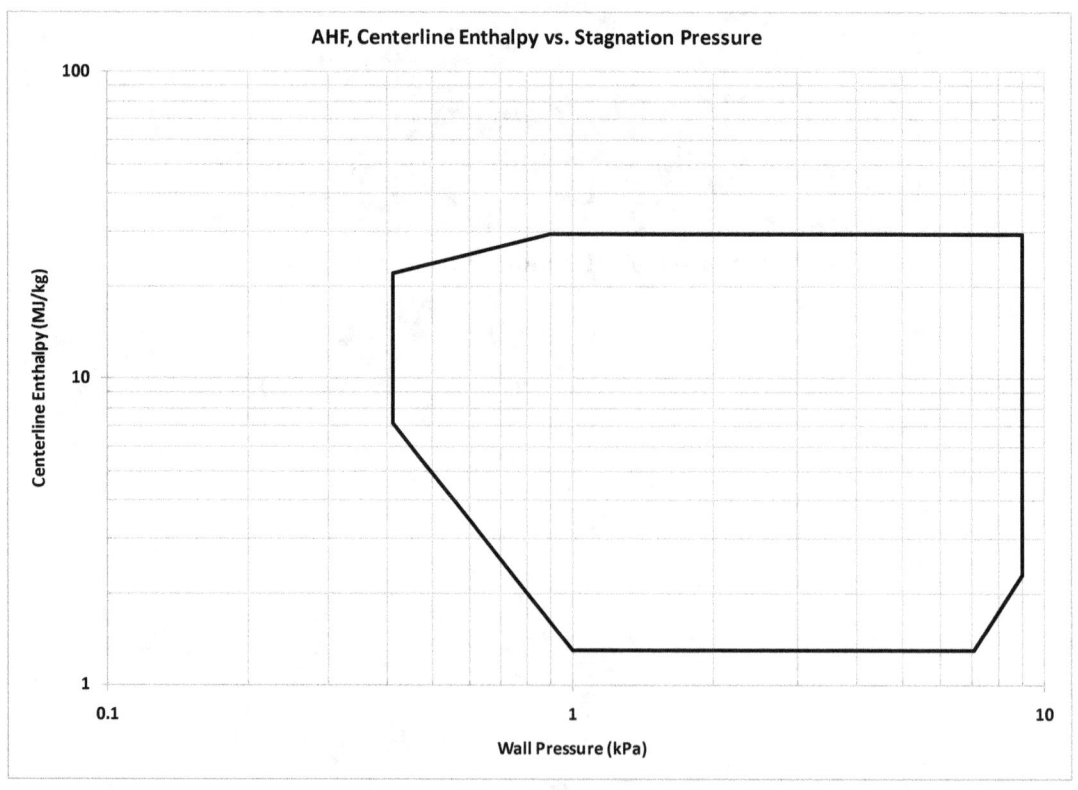

Figure F-13. AHF Stagnation Performance, Centerline Enthalpy vs. Pressure.

Panel Test Facility (PTF)

The PTF uses a segmented arc heater similar to IHF and AHF to drive the facility and operates off the 20-MW power supply. This facility is specifically designed to perform panel shear type or protuberance type testing and is coupled with a semi-elliptic nozzle. Two semi-elliptic nozzle sizes are available, a 4 in. by 17 in. and a 1.5 in. by 6.7 in., which result in either a 14-in by 14-in panel test size or a 4-in by 4-in panel test size. The smaller nozzle and resulting smaller panel test size result in higher surface pressures, heat flux, and shear on the test sample; however, these conditions are not shown in Table F-1. A picture of the larger nozzle and a test sample are shown in Figure F-14. The test sample can be inclined in the flow from -4 to +8 degrees and can be changed during the run. Run times of up to 30 minutes are possible. There is good optical access to the test model face during the run.

The heater operates at chamber pressures from 1 to 9 atm and bulk enthalpies from 3,000 to 15,000 BTU/lbm. Convective heating rates on the model surface range from 0.5 to 30 BTU/ft^2-sec. Pressures range from 0.0006 to 0.05 atm. Facility performance is presented in Figure F-15 through Figure F-17.

Figure F-14. Panel Test Facility Test Cabin showing Semi-Elliptic Nozzle and Test Panel.

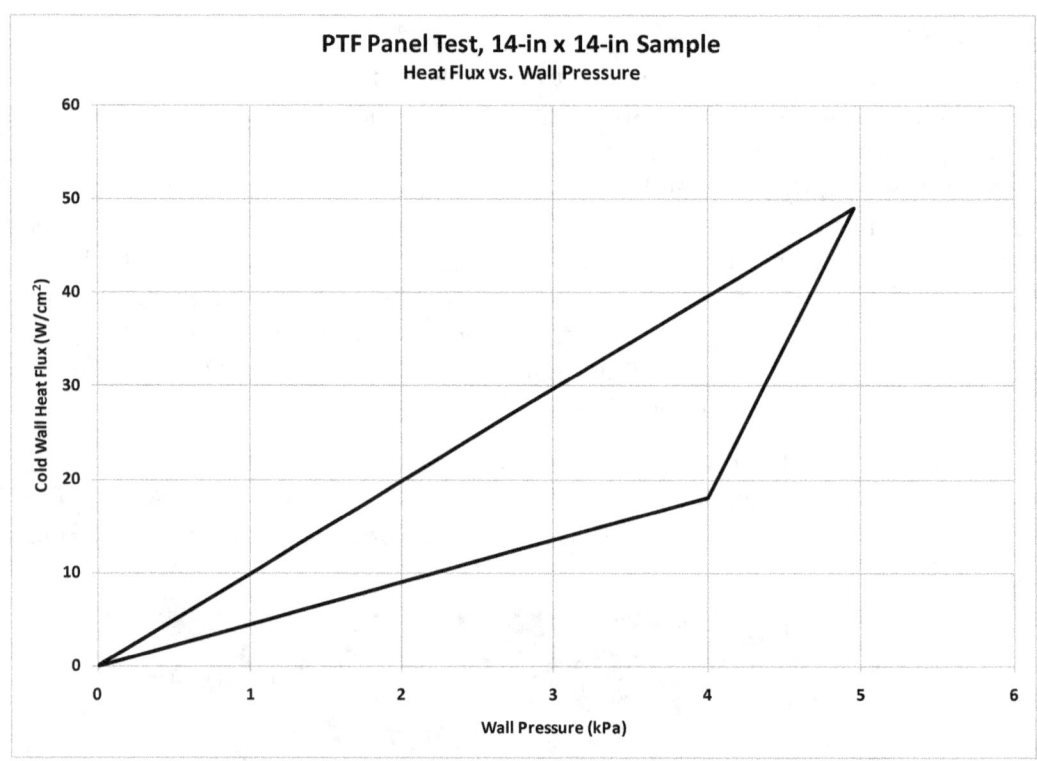

Figure F-15. PTF Panel Test Performance on a 24-in x 24-in Sample, Heat Flux vs. Wall Pressure.

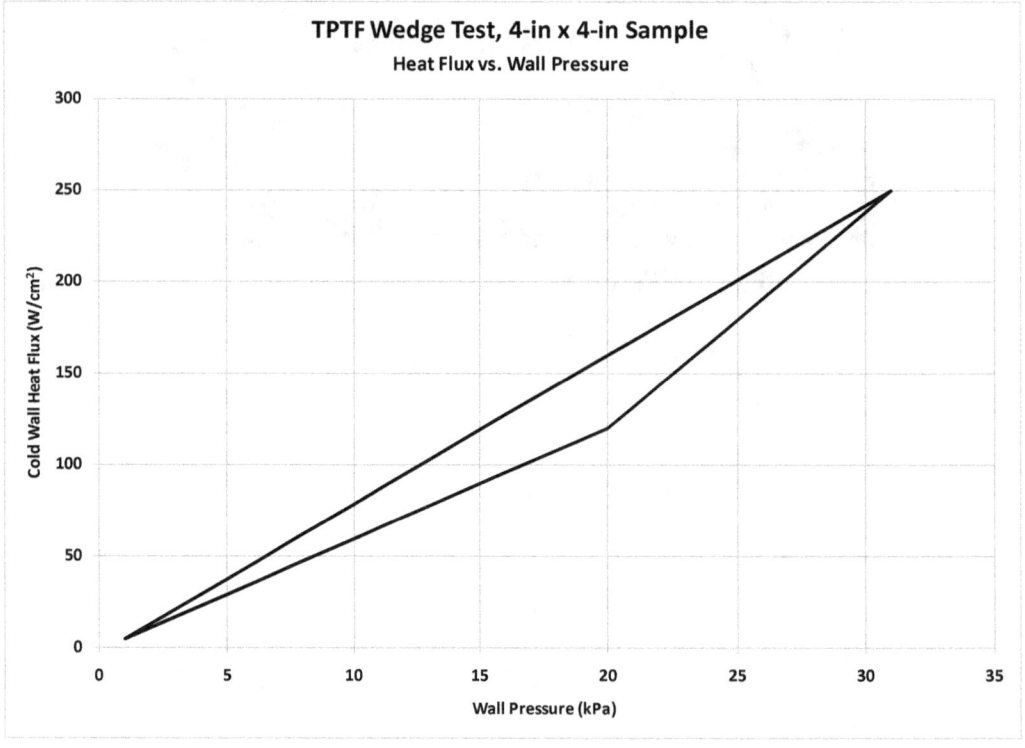

Figure F-16. TPTF Panel Test Performance on a 4-in x 4-in Sample, Heat Flux vs. Wall Pressure.

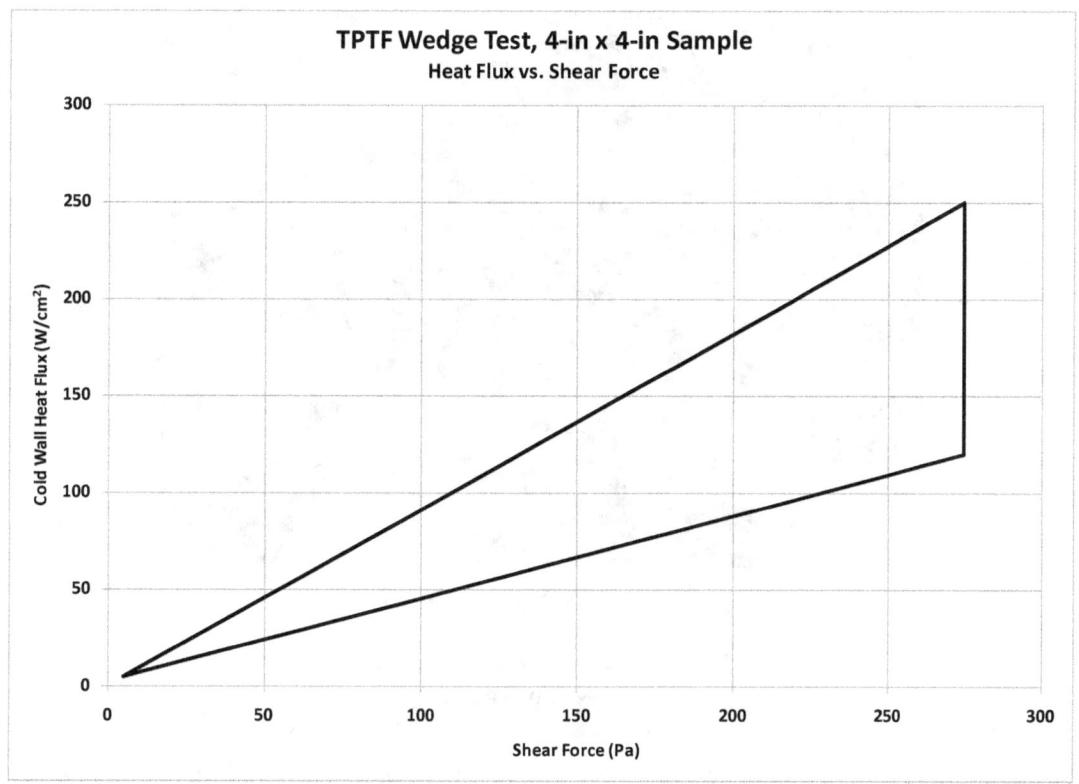

Figure F-17. TPTF Test Performance on a 4-in x 4-in Sample, Heat Flux vs. Shear Force.

Turbulent Flow Duct (TFD)

The TFD uses a Huels heater to drive the facility and operates off the 20-MW power supply. The heater can operate on either air or nitrogen. This facility is specifically designed to perform shear type testing and has a 2-in. by 9-in. channel flow test configuration as shown in Figure F-18. The test sample is installed so it is part of the channel wall and can be subjected to higher shear values than the other ARC facilities. Test sample sizes can be either 8 in by 10 in. or 8 in. by 20 in.

Bulk enthalpies range from 1,300 to 4,000 BTU/lbm with convective heating rates on the model surface from 2 to 60 BTU/ft^2-sec and pressures from 0.02 to 0.15 atm. Model shear force values range from 1 to 15 lb/ft^2.

Figure F-18. Turbulent Flow Duct.

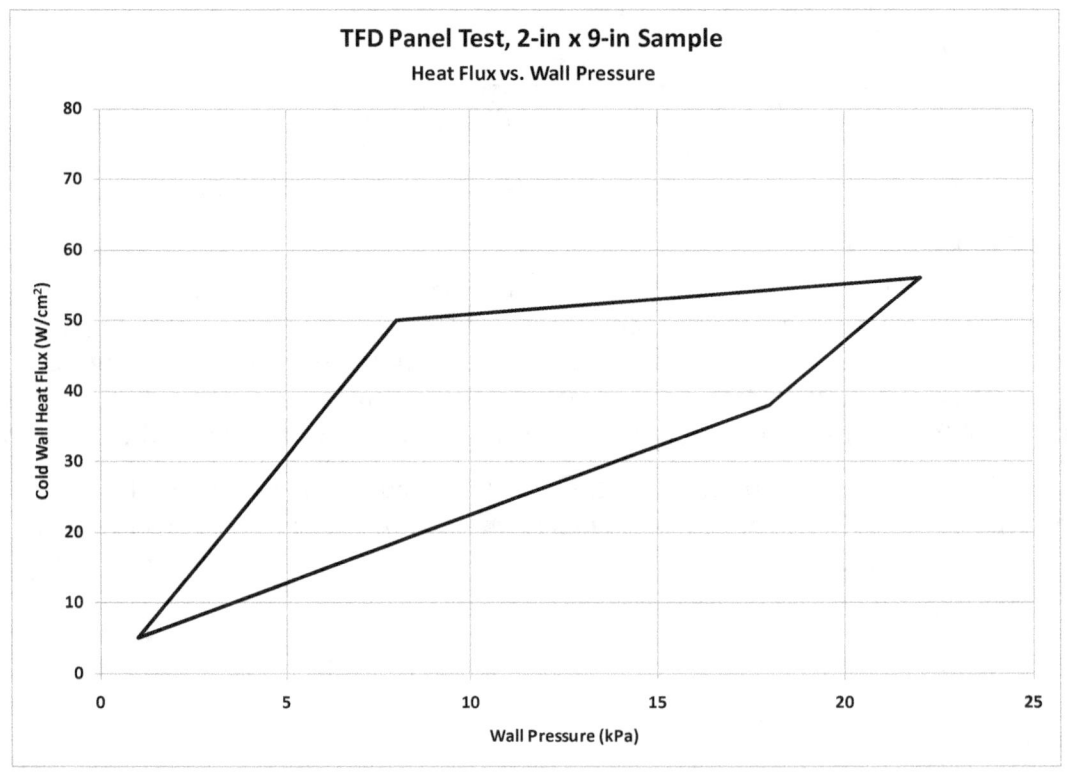

Figure F-19. TFD Performance on a 2-in x 9-in Sample, Heat Flux vs. Wall Pressure.

Johnson Space Center Arc Jet Complex

The JSC complex contains two available test bays. Both have vacuum exhaust capability and are occupied with active facilities, TP-1 and TP-2. Both test bays share a common power supply, steam-ejector driven vacuum system, water cooling system, high-pressure gas system, and other auxiliary systems. The power supply is rated for 10-MW continuous power output; however, both facilities typically operate at power levels around 5 to 6 MW.

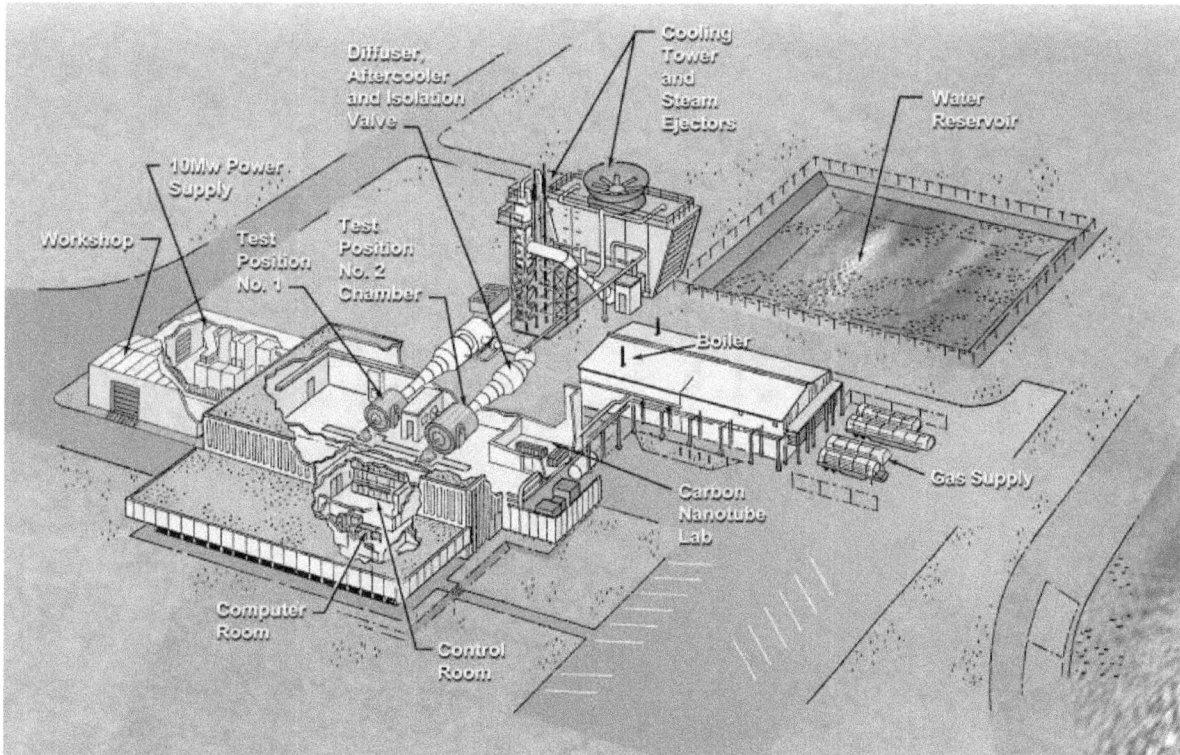

Figure F-20. Sketch of the JSC Arc Jet Complex.

Both of the arc heaters that drive TP-1 and TP-2 are segmented heaters with a tungsten button cathode on the upstream end and a conical copper anode on the downstream end. The test gas is a mixture of nitrogen and oxygen with the nitrogen and oxygen being individually injected at discrete locations and mixed along the heater bore. The percentages of nitrogen and oxygen can be varied. The heater can also be operated on pure nitrogen. Argon is not injected over the electrodes during testing, so only oxygen and nitrogen are injected into the heater as the test gas.

Figure F-21. JSC Arc Jet Facility Showing TP-1 on the Left and TP-2 on the Right.

Test Position-1 (TP-1)

The TP-1 facility is a channel-flow facility that is unique to the Agency. The facility is designed with a channel flow configuration having a 2-in. wide channel and three different test locations along the 10-degree half-angle nozzle as shown in Figure F-22. The test sizes are: 8 in. by 10 in., 12 in. by 12 in., and 24 in. by 24 in. The test sample is installed so it becomes part of the nozzle side wall. The pressures and heat flux are highest at the upstream test position and lowest at the downstream test position. Test performance for the 12-in. by 12-in. and 24-in. by 24-in. sample sizes are shown in Figure F-23. A new 4-in. by 4-in, 10-degree half-angle duct has been fabricated that will provide higher heat flux, pressure, and shear than the 8-in. by 10-in. test station.

The opposing side wall facing the test sample can be instrumented with pressure and heat flux gages, or it is possible to heat the wall to provide a radiant heating capability. Optical access to the test sample is not possible.

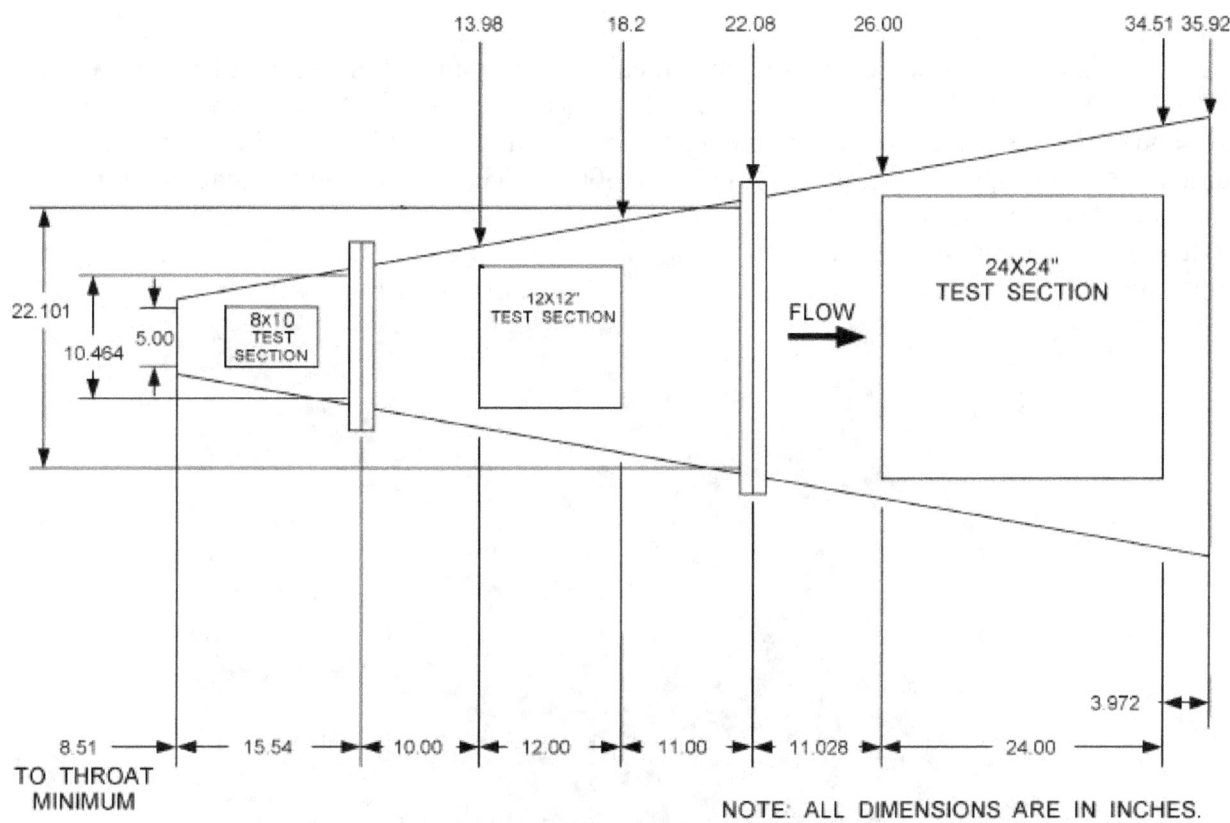

Figure F-22. TP-1 Channel Nozzle Showing Three Test Locations.

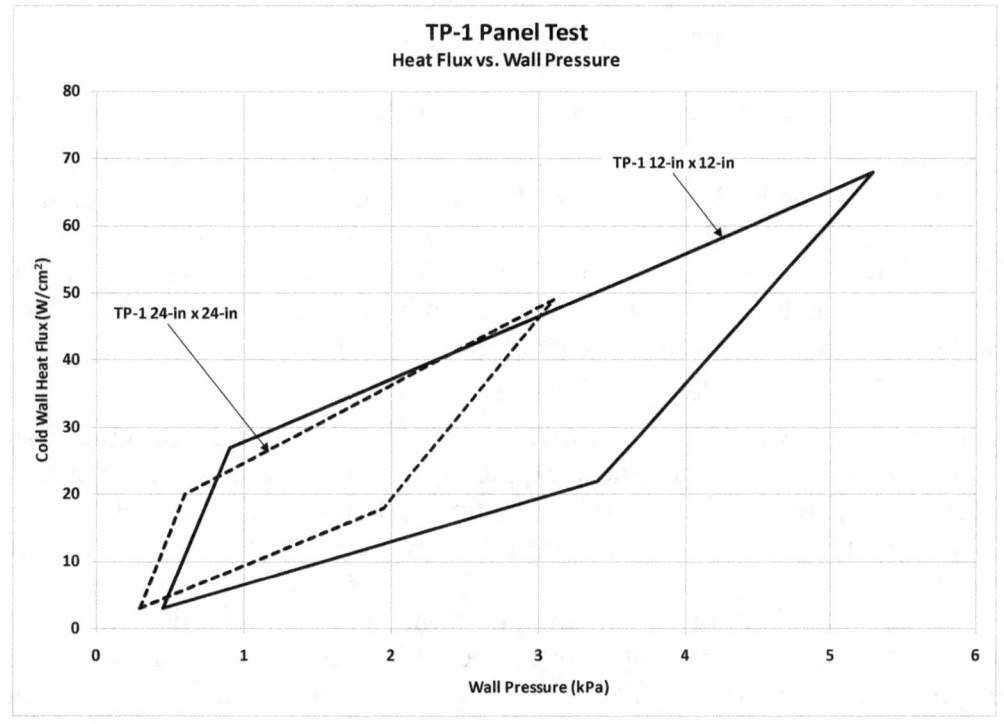

Figure F-23. TP-1 Panel Test Performance for Heat Flux vs. Wall Pressure.

Test Position-2 (TP-2)

TP-2 uses a family of 15-degree half-angle conical nozzles with exit diameters of 5, 7.5, 10, 15, 20, 25, 30, 35, and 40 in. (shown in Figure F-24). The 12-foot diameter test chamber allows video and optical temperature measurement access. Two hydraulically controlled sting arms are available that can support models weighing up to 500 lbs. The facility can test stagnation type models or shear type models using a wedge configuration. The model size is dependent on the nozzle exit diameter being used, but four standard wedge configurations are available with test sample sizes of 4.5 in. by 5 in., 6 in. by 6 in., 12 in. by 12 in., and 24 in. by 24 in.

Figure F-24. TP-2 40-in. exit Diameter Nozzle Configuration.

Arnold Engineering Development Center Arc Jet Complex

The AEDC arc jet complex contains four free-jet to atmosphere test bays, of which two are occupied with active arc jet facilities—H1 and H3. In addition, AEDC has a vacuum-capable test bay, which is occupied by the H2 facility. The two free-jet to atmosphere facilities, H1 and H3, provide high-heating and high-pressure simulation, which is of little interest for NASA missions. Therefore, the H1 and H3 heaters will not be evaluated in this report. The H2 heater, however, does have test capabilities that overlap an area of interest for NASA, specifically, test simulation in higher pressure, higher-shear operating ranges.

The H2 facility is driven by a Huels arc heater, as shown in Figure F-25, from the 70-MW power supply and uses air as the test gas. The heater typically operates at a maximum power of 42 MW. A five-sting arm, computer controlled model injection system can inject models into the flow at variable speeds and move models axially in the test flow during the test. Multiple models or probes can be put on each sting arm, as shown in Figure F-26. Stagnation type models or wedge type models can be tested. Maximum run times are from 3 to 30 minutes depending on the test condition. The nozzle exit diameters range from 5 to 24 in. Flow-field stagnation enthalpies range from 1,200 to 6,500 BTU/lbm. Facility performance is shown in Figure F-27 and Figure F-28.

Figure F-25. AEDC H2 Huels Type Arc Heater.

Figure F-26. View of AEDC H2 facility Test Cabin and Model Injection System.

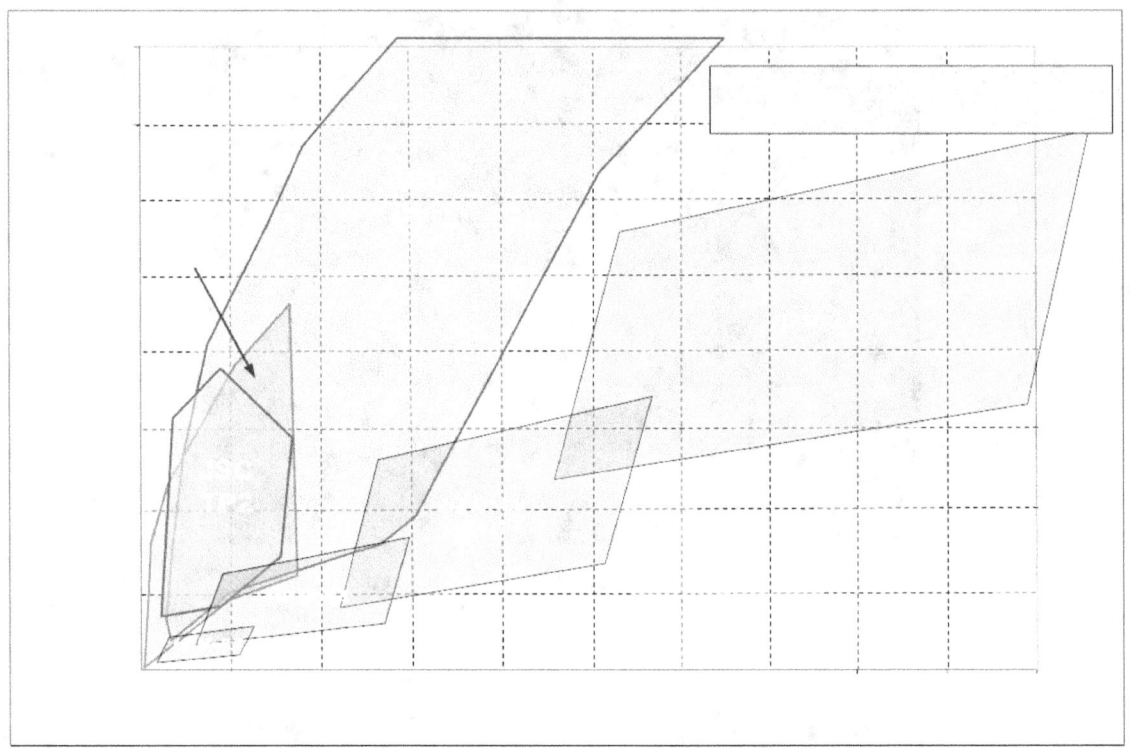

Figure F-27. H2 Wedge Test Performance Compared with NASA Wedge Test Performance.

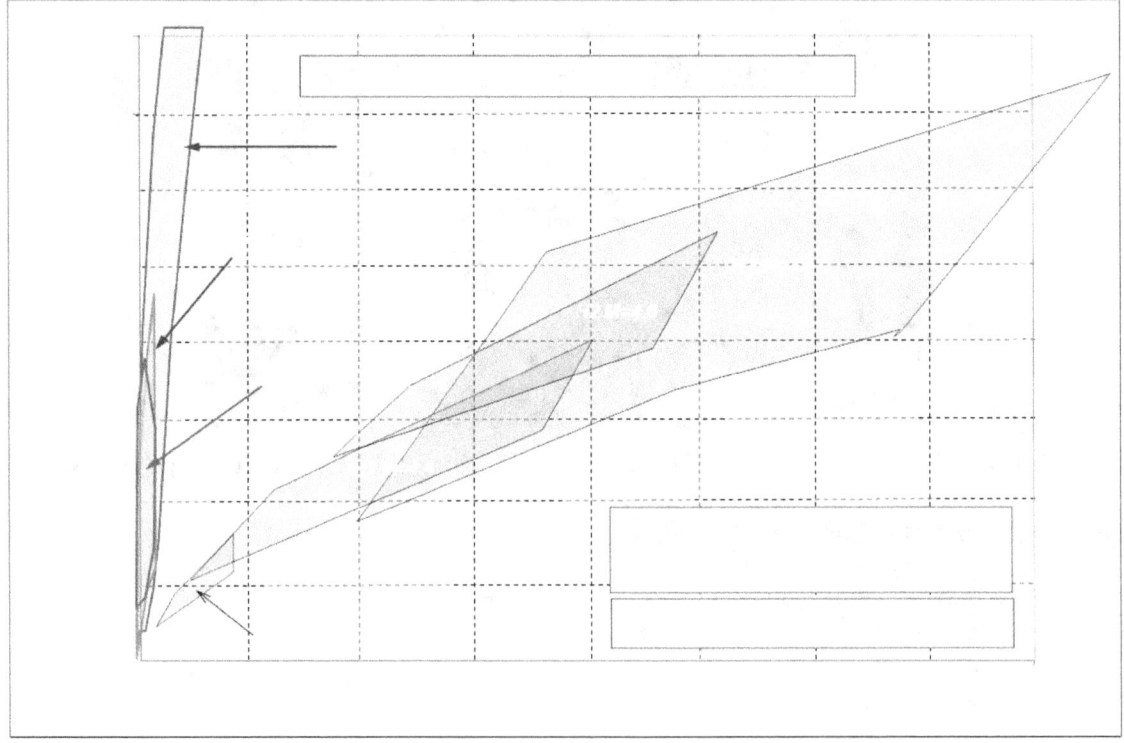

Figure F-28. H2 Stagnation Test Performance Compared with NASA Facility Performance.

Boeing LCAT Arc Jet Complex

The Boeing arc jet complex has two test bays, one free-jet to atmosphere and one with vacuum exhaust capability. The LCAT facility occupies the vacuum capable test bay and is driven by a 12-MW power supply. The LCAT facility has a Huels arc heater. A rotary model injection system can test up to four models per run; however, typically only three models are tested per run. Three conical nozzles are available with 4-in., 6-in., and 12-in. exit diameters. In addition, the LCAT facility has a square nozzle and a semi-elliptic nozzle. Performance of the LCAT facility is shown in Figure F-31 through Figure F-34.

Figure F-29. View of Boeing LCAT Facility.

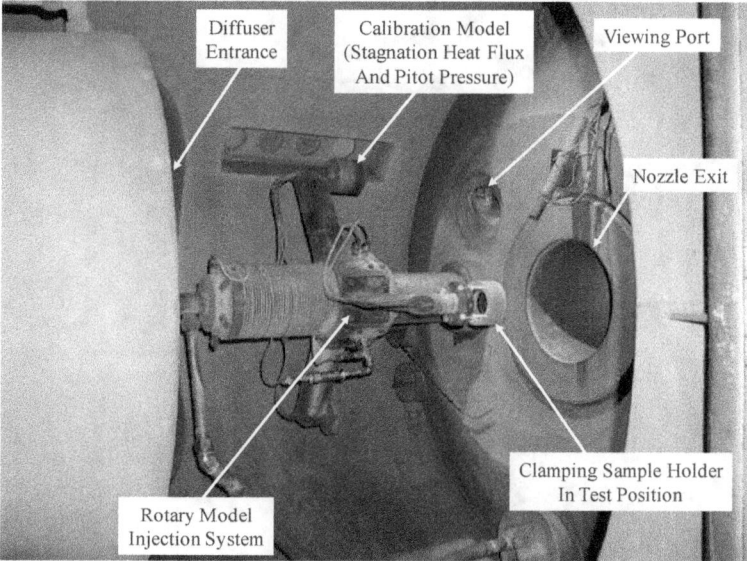

Figure F-30. View of LCAT Test Cabin Interior and Model Injection System.

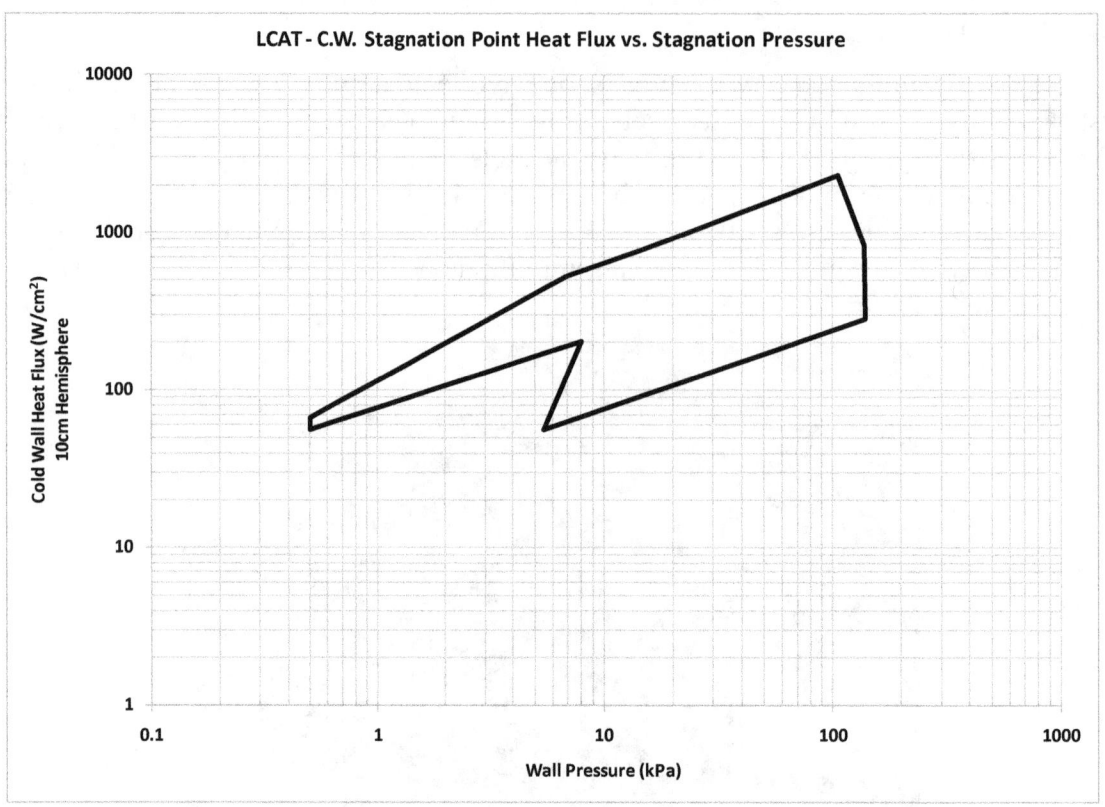

Figure F-31. LCAT Stagnation Performance of Heat Flux vs. Pressure.

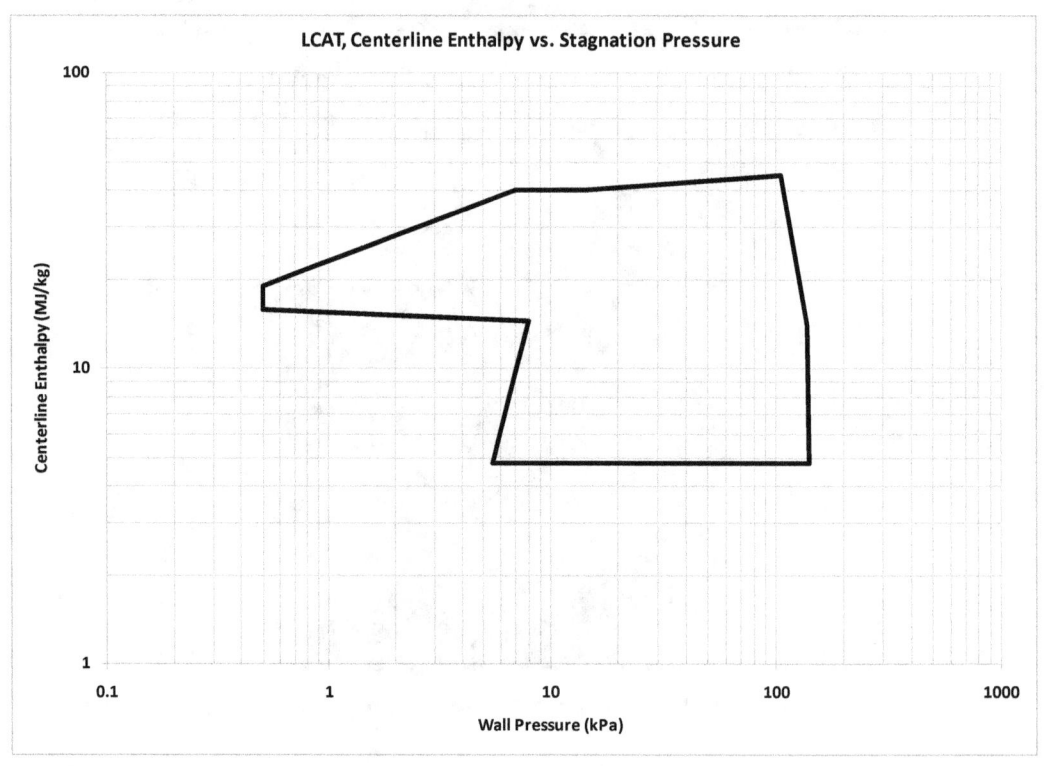

Figure F-32. LCAT Stagnation Performance of Centerline Enthalpy vs. Pressure.

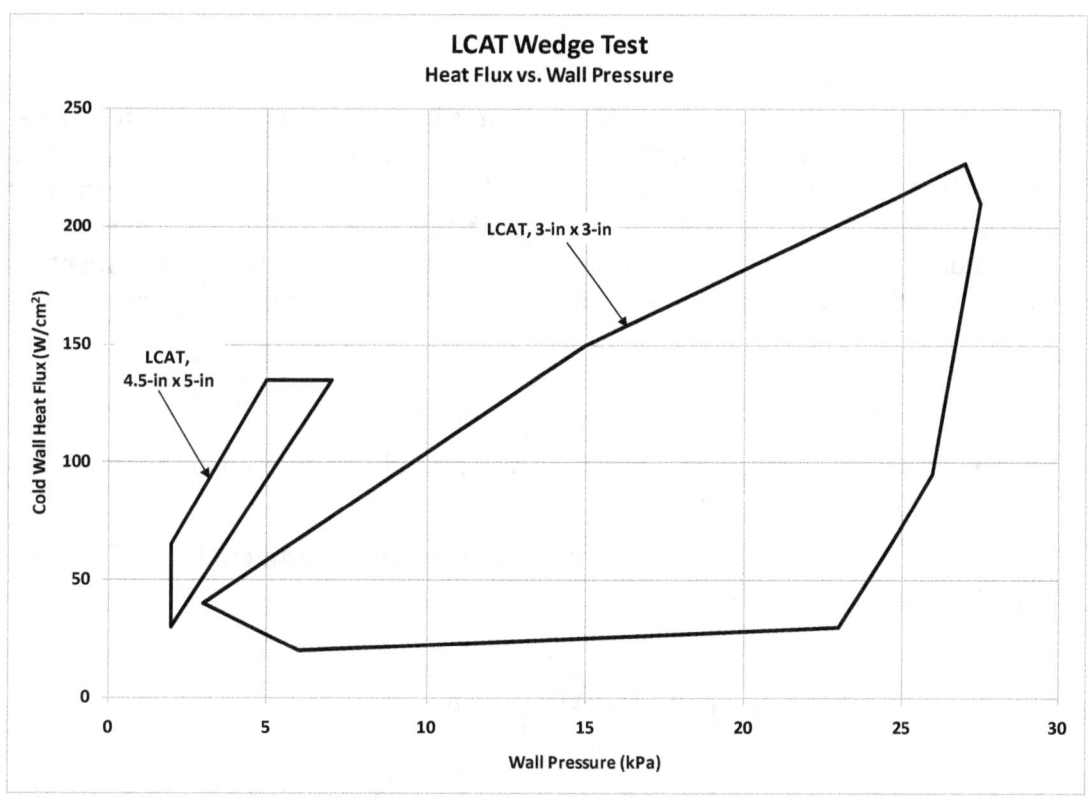

Figure F-33. LCAT Wedge Test Performance, Heat Flux vs. Wall Pressure.

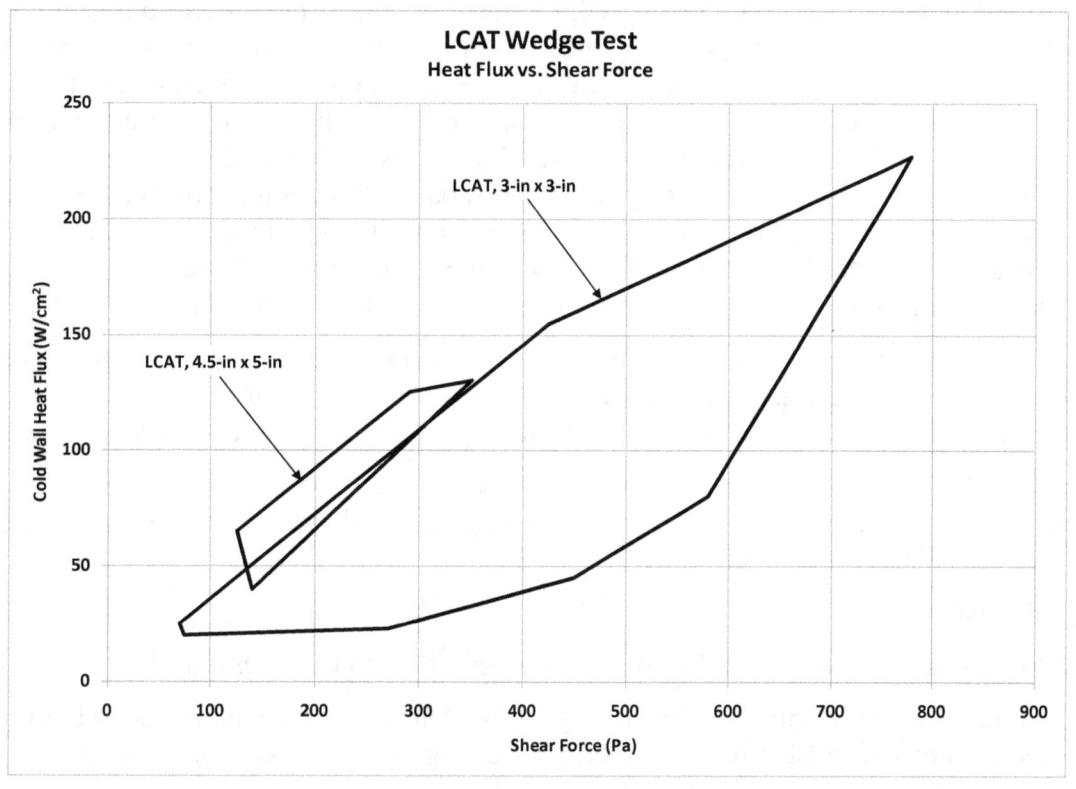

Figure F-34. LCAT Wedge Test Performance, Heat Flux vs. Shear Force.

Appendix G: Investment Options

Several options were evaluated for NASA's 30-year investment in arc jets to support TPS development and operation for future missions. They were compared on the basis of capability, capacity, availability, and cost. The AJEWG assumed that no more than one new construction complex would be identified as a viable funding possibility. The basis for estimated costs over 30 years is included in the discussion of each option. A table of individual cost elements for each option is included at the end of the section [deleted from public version of report].

The options identified for evaluation are:

- Close the existing arc jet complexes at ARC and JSC
- Continue operating both arc jet complexes, but provide no additional investment for revitalization and recapitalization
- Make both ARC and JSC complexes healthy, including investment for revitalization and recapitalization
- Close the ARC complex and make the JSC complex healthy
- Close the JSC complex and make the ARC complex healthy

1.1 Close the Existing Arc Jet Complexes at JSC and ARC

The AJEWG recognizes that arc jet tests will be required to support future missions, because it is not anticipated that missions beyond LEO will be flown without arc jet tests, so this option means that test services would be procured outside NASA. The only available Government facility is at AEDC, and its low enthalpy, high-pressure conditions are not a good match for the environments anticipated for most NASA missions. The only available commercial facility is LCAT, at Boeing, but its capabilities do not cover the full test range needed. International facilities, most notably Sirocco in Italy, appear to be capable, but complications associated with data security and ITAR regulations suggest that regular international testing is untenable. The Air Force has relied on NASA test capabilities for some strategic vehicles and stated they would not test materials in a foreign facility, even if ITAR restrictions were not an issue.

Although the near-term cost savings may seem attractive, the long-term costs of relying on an external facility have a high risk, and may ultimately exceed the cost of maintaining a NASA capability. Also, the availability of arc jet test time, especially for rapid access in a crisis, would be poor.

Most importantly, the technical knowledge and skills associated with arc jet design, diagnostics, and operations will be lost to NASA.

Estimated Costs

The cost to close and dispose of both complexes is estimated to be several million.

Although at face value this option offers savings, this option is unacceptable due to high risk of lack of access when needed and the loss of critical testing capability and knowledge.

1.2 Continue Operating Both Arc Jet Complexes But Provide No Additional Investment for Revitalization and Recapitalization

Both complexes could continue to operate for several years without major infrastructure investment. The long-term deferral of refurbishment, which is a decision to run to failure, effectively amounts to a decision to close the complexes at an uncertain time, when a serious failure will render the complexes inoperable.

Estimated Costs

Two cases are evaluated for this option: the continuation at the current level of operation and operations at a reduced level.

Reducing the operations at ARC will result in a decrease to maintenance, personnel, and capacity, and will essentially become proficiency operations only. Operations at JSC will decrease from two shifts to one shift.

These options are also unacceptable, since continuing to operate the facilities without a strategy to upgrade or replace to support the long-term needs of NASA will eventually result in an unplanned loss of capability.

1.3 Make Both ARC and JSC Complexes Healthy, Including Investment for Revitalization and Recapitalization

In recent years, the facilities at ARC and JSC have been heavily utilized, with ARC handling all planetary mission testing for the Science Mission Directorate and JSC handling the bulk of Shuttle testing. ARC has provided backup testing for the Shuttle, and there has been considerable coordination of testing between the two facilities for Orion support, both through the TPS Advanced Development Project and subsequent Orion insight and oversight work.

Operational benefits associated with the existence of two geographically distinct facilities have been demonstrated in recent years. The redundancy of some capability can compensate for a lack of reliability at either complex. The combined capacity permits robust response to surges in test requirements, which has been particularly helpful for real-time support of Shuttle missions.

Interaction of the personnel at different facilities is also fruitful. The distinct groups have different approaches to dealing with similar technical issues, so more options are proposed, and their relative merits are vigorously assessed. At times when mission test requirements are low, the excess capacity can be used to investigate differences in the two facilities and the implications for differences in test outcomes. Members of both teams can develop deeper understanding of arc jet design and operations through the unique features of each facility.

Ultimately, this option should ensure current test capabilities and maintain a robust pool of highly skilled personnel for arc jet operations. In the short term, vibrant teams at two locations are likely to promote innovative approaches for facility development, but in the long term, the cost of maintaining two complexes with separate operating teams will leave little funding to implement innovations.

Estimated Costs

This option is not recommended because it does not adequately address NASA's long-term mission goals. Unfortunately, the benefits of operational robustness and personnel interaction have a considerable cost due to maintenance of two sets of infrastructure and high staffing levels.

The outdated technology at both facilities would be retained, so that the existing capability would be delivered inefficiently.

Some of the redundancy benefits of distinct facilities can be realized at a single complex through judicious design of infrastructure and arc heaters. Similarly, some of the benefits of two distinct teams cross-pollinating to improve test practices can be realized at a single location if personnel are organized in separate groups. A single complex might have a group concentrating on testing to support mission design and operations, while a second team is devoted to material and TPS concept development. The groups could coordinate facility usage and share ideas on potential improvements. Such an approach has been effectively employed at AEDC.

1.4 Close the ARC Complex and Make the JSC Complex Healthy

The complex at JSC has less capability and capacity than the ARC complex and, consequently, has somewhat lower operating costs. It is capable of supporting LEO missions well, but its lower power levels provide limited utility for more energetic entry profiles. This option is somewhat better than closing both facilities, because some arc jet expertise is retained and small samples can be tested effectively, but it does not adequately address NASA's long-term mission goals.

Estimated Costs

This option is unacceptable in the long term, since it leaves NASA without the higher performance capability that currently exists at ARC.

1.5 Close the JSC Complex and Make the ARC Complex Healthy

Closing the JSC arc jet complex and making the ARC complex healthy has the potential to reduce total arc jet costs by decreasing capacity and backup robustness without giving up capability, because JSC operating conditions are a subset of those achievable at ARC. JSC currently has superior functionality in its ability to vary gas mixtures and avoid argon in the test stream, but this capability can be implemented at ARC at a modest development cost.

The cost reductions that would accrue through closure of one facility are likely to be smaller than anticipated. JSC also operates the RHTF with the same crew now operating the ARMSEF. Either the radiant capability would also be lost, or the personnel reductions would be smaller than the total number currently operating the arc jets. The remaining team would likely be less productive than they are currently. Furthermore, the lack of a second operation may reduce the impetus to deliver efficiency improvements at the remaining facility. The cost savings that are achieved are unlikely to be redirected to arc jet support at a different Center, making it less likely that the remaining complex can upgrade its capability. Existing technology will be maintained, but the pool of talent that can drive improvement will shrink.

Estimated Costs

This option should reduce costs relative to operating two facilities, but it reduces capacity and introduces some availability risk without delivering substantial capability improvement. In the long term, if the Agency chooses not to improve capability, this is a viable option. But it is not recommended because it does not adequately address NASA's long-term mission goals. There are significant human resource risks associated with closure of one facility. The TPS communities at ARC and JSC have worked together extremely well in the years since the

Columbia accident, displaying unity of purpose through RTF and into Orion TPS development and oversight. Facility closure is likely to lead to Center-centric decision making.

Finally, there are real risks associated with concentrating resources at ARC. The likelihood of major damage from earthquakes is small but real, and the consequences for Agency mission support could be severe. The cost of upgrades that comply with structural codes is relatively large in California. ARC might soon require a waiver for air quality. A new boiler would address that issue, but air and water quality regulation limits will continue to become more stringent in California, and the cost of compliance could be high. Finally, ARC is approaching its quota for low-cost hydroelectric power, and the cost and availability of electrical power might be problematic in future decades.

Appendix H: Work Statement

Evaluation Purpose

Provide engineering support to the NASA Office of the Chief Engineer for, and be a member of, the Arc Jet Evaluation Working Group (AJEWG), a focused, independent evaluation of NASA arc jet facilities, considering the availability and use of other hypersonic thermal testing facilities, against planned and future mission requirements and provide findings that will help inform possible long-term investment strategy alternatives for NASA's arc jet capability.

Introduction

An arc jet facility provides ground-based, long-duration, high-energy density (enthalpy) environments for testing thermal protection materials. These facilities are designed to simulate heating and aerothermal forces experienced during hypersonic deceleration of a spacecraft through a planetary atmosphere. The environments the arc jet can achieve allow material engineers and spacecraft designers to: develop new, innovative thermal protection materials; select the appropriate heat shield, backshell, and seal design, and validate estimates of its response; and flight certify and establish required performance margin for the thermal protection system (TPS) design during entry, descent, and landing (EDL). They also provide NASA the capability to test, analyze, and consider repair alternatives to on-orbit thermal protection system anomalies, as well as the capability to support mishap investigation boards.

In operation, high-pressure air is heated electrically to a flight-like energy enthalpy. This high-energy air is expanded and accelerated to hypersonic speeds through a nozzle into a vacuum test chamber where the test article(s) is/are exposed to a flight-relevant heating, pressure, and shear environment. The test article is inserted into the flow for periods of up to an hour to simulate the long duration exposure of an actual entry. Data is recorded continuously from model-embedded sensors and optical instrumentation. Test articles include ablative, as well as reusable materials and systems.

The arc jet facility includes elements that are required to accomplish given program or project requirements, as well as to maintain the facilities in ready-to-operate condition. This includes the skilled workforce, support facilities/equipment, suppliers/supplies, and contracts that, when combined, perform the arc jet function. Facilities and equipment include all real property/structures such as buildings, software and business systems, and any personal property that might be required to operate the arc jet. The workforce includes the technical staff (both civil servant and contractor) required to safely and reliably operate the facility, as well as to conduct analysis, engineering and science, and produce reports. The workforce includes all support staff required to conduct business-like functions such as scheduling and budget formulation, execution, and reconciliation. Suppliers/supplies, contracts, and other items for consideration include O&M, data/data storage, knowledge (intellectual property, reports, and collective undocumented wisdom and experience), old and current contracts and contract-related termination liability, and all other ancillary supplies, tools, utilities, and consumables required for safe arc jet operation.

Charter

The AJEWG will document requirements and capabilities for current and planned NASA missions, as well as aeronautics research milestones and advanced entry system technologies that drive the utilization of arc jet facilities used for hypersonic thermal testing. Those requirements will be evaluated against the performance envelopes and throughput capacities that are supported in the current NASA thermal protection system test facility configuration. Support for all mission phases, from precursor technology maturation through sustaining engineering of operational systems, should be considered. Driving thermal testing requirements should consider human and robotic Mars entry; human and robotic Earth return from the LEO, the Moon or Mars; deep space robotic Earth return; robotic science missions to Venus and the Giant Planets; and atmospheric entry of robotic science missions to Titan. Table H-1 is provided as a guide for one method that could be use to identify driving thermal testing requirements for planned and future missions. Hypersonic thermal test facility throughput capacities should consider use of all available resources (both NASA and non-NASA), maximum throughput capacity of those resources, and envelope of operation of those resources against thermal testing requirements for future missions. Throughput capacities of test facilities are not expected to be stagnant, as consideration should be given for planned maintenance and facility upgrades. A report will be generated to capture results of the evaluation and based on findings and analysis; the report will include possible investment strategy alternatives for NASA's hypersonic thermal test facility capability.

The AJEWG will coordinate their efforts and findings with the Technical Capabilities, Requirements, and Utilization Panels of the Facilities Program Board, to inform these panels as they develop a 5-10 year Agency investment strategy. Findings should consider:

- Maintaining current capabilities for a specified period of time,
- Consolidation and upgrade of current capabilities as deemed appropriate,
- New construction ("Greenfield" approach),
- Other categories to include combinations of alternatives and utilization of outside capability such as Arnold Engineering Development Center (AEDC) or Boeing St. Louis arc jets.

Table H-1: Program and Project Requirements for Arc Jet Capability

Program	Gas	Input Power (MW)	Typical Test Time (heat load)	Types of Test Article	Model Test Positions	Nozzle Exit (Inches)	Mach Number	Flow Enthalpy (Btu/lb$_m$)	Surface Pressure (atm)	Heating Rate (Btu/ft^2-sec)
*Human Mars Entry										
*Robotic Mars Entry										
*Human Lunar Return										
Human ISS Return										
Robotic Lunar Return										
*Human Mars Return										
*Robotic Mars Return										
*Deep Space Robotic Return										
Robotic Venus Aerocapture										
Robotic Venus Entry										
Robotic Giant Planet Aerocapture										
Robotic Giant Planet Entry										
Robotic Titan Aerocapture										
Robotic Titian Entry										
*ISS Down Mass										

*Most likely future missions.

Organization

The AJEWG will be comprised of knowledgeable individuals that do not have a stake in the outcome, and this working group will be led by Mike Ryschkewitsch, NASA Chief Engineer. The AJEWG Lead will engage technical experts and advisors as required.

Deliverables

- Draft Report: February 24, 2010, including possible investment roadmap alternatives (rough order estimated cost, schedule and risk) that inform the 5-10 year infrastructure investment strategy to provide capability through 2040.

- Final Report: March 31, 2010.